THE
WONDERS
OF THE
WILD PLACES

THE WONDERS OF THE WILD PLACES

RACHEL LISTER JONES

First published 2022

Copyright © Rachel Lister Jones 2022

The right of Rachel Lister Jones to be identified as the author of this work has been asserted in accordance with the Copyright, Designs & Patents Act 1988.

All rights reserved. No part of this book may be reproduced, stored in a retrieval system, or transmitted in any form or by any means, electronic, electrostatic, magnetic tape, mechanical, photocopying, recording or otherwise, without the written permission of the copyright holder.

Published under licence by Brown Dog Books and
The Self-Publishing Partnership Ltd, 10b Greenway Farm, Bath Rd,
Wick, nr. Bath BS30 5RL

www.selfpublishingpartnership.co.uk

ISBN printed book: 978-1-83952-533-9
ISBN e-book: 978-1-83952-534-6

Cover design by Andrew Prescott
Internal design by Andrew Easton
Illustrations by Hannah Doyle

Printed and bound in the UK

This book is printed on FSC certified paper

In memory of my Grandpa and Aunty, Bill and Daphne Lister, who encouraged me to write, and to my parents, Cindy and Peter Jones who passed on their love of nature to me.

Contents

Introduction	9
If you go down to the woods today	19
Hedges and edges	47
On the farm	67
Up high and heaths	93
Simply messing about in boats	123
The deep blue sea	165
Urban Jungle	239
Hope for the Future	299
Appendix	305
Sources	312
Bibliography and References	345
About the author	352

Introduction

For centuries, human beings believed that the natural environment was just something to be exploited – either by hunting animals or eating plants and fungi; using wood to build houses and to keep warm, utilizing animals and plants to make clothes, and curing ailments using medicinal plants. Then, starting with the Renaissance in the 16th century and continuing during the Age of Enlightenment in the 18th century there was a sea-change in how people perceived nature. As the environmental historian Dr Rob Lambert describes it, there was a shift, "... from use to delight." During the Renaissance the first books about nature were printed. Through the Age of Enlightenment people began to understand the natural world more from a scientific point of view, with people like Carl Linnaeus categorising the natural world. Although the binomial nomenclature had been invented a century earlier, it was Linnaeus who had the organisational skills and diligence to create the system we still use today to name every living organism. However, at the same time people appreciated the complexity and the beauty of nature. The Romantic poet Samuel Taylor Coleridge was one of the people who took inspiration from nature. He said, "Everything has a life of its own...and...we are all one life." This was a view reflected by his fellow Romantic poets.

However, during the Industrial Revolution there was a seismic shift in how people lived. After generations of living on the land, much of the population moved to urban areas. This was the start of our disconnection with nature. People could no longer hear the songbirds singing as they toiled the land or see the changing of the seasons, they now lived and worked in dark, noisy, cramped conditions. However, the new city-dwellers did not totally forget about what they had left behind. In late Victorian times people's appreciation of nature grew, which led to strange crazes, such as pteridomania (fern collecting) and orchidelirium (orchid collecting), and organisations such as the R.S.P.B. and the National Trust being set up to protect the natural environment. However, at the same time, Victorian explorers visited the remotest corners of the world plundering and hunting in order that people could see things they had never seen before in natural history museums and the wealthy could have exotic plants growing in their new glasshouses.

Since the Second World War, and particularly in the last 50 years, there has been a move away from appreciating and protecting the environment, and a return to the exploitation of the natural environment on a catastrophic scale. This sea-change has happened not just in the United Kingdom but world-wide. Increased industrialisation has led to more carbon dioxide, nitrogen, and other greenhouse gasses in the atmosphere. This has caused global temperatures to rise, causing glaciers and the ice caps, especially the Arctic, to start melting. Mechanised intensive agricultural practices, especially the increased use of synthetic fertilizers and pesticides, has led to a dramatic decline in bird and insect populations, and has damaged the land itself making it less fertile. Population growth has led to a greater pressure on land – using it for housing and growing food rather than preserving it for nature. All over the world deforestation, failure to maintain and care for grasslands and heathland, erosion, and the spread of invasive species has led to loss of habitat and tougher conditions for flora and fauna in which to survive. Some species have been able to adapt, but the change has been so great that many have not been able to. Climate change has also led increasingly to more severe weather conditions. In 2020 wildfires in Australia killed an estimated one billion animals, some of which can only be seen on that continent. Rapid deforestation of the Amazon and rainforests in Borneo and Madagascar might have destroyed plants, birds, and animals not even known to science. Some of the species that we do know about are not only important from a biodiversity point of view but can also be used to treat human diseases. For example, medicine made from a species of periwinkle found in Madagascar is used to treat various forms of cancer. In addition, failure to manage our waste products properly has led to it being dumped in the sea. Combined with overfishing this has led to many problems with our seas. Increasing temperatures are making our oceans more acidic, damaging vital habitats, such as coral reefs. Although nationally and globally governments are promising to cut carbon emissions, halt deforestation, and improve habitats, currently there seems to be more words than action and eminent naturalists, including Sir

The Wonders of the Wild Places

David Attenborough, have warned that we must take action this decade to avert a catastrophe. However, a recent United Nations report warns that rather than cutting emissions by 45% by 2030, which is vital to avert a rise of 1.5°C, on current trends emissions are due to increase by 16%, which will have a devastating effect on all living things on this planet including us.

For the last fifty years for many people in the United Kingdom our natural environment and the plants, birds, insects, and animals which also inhabit this island have just been regarded as something in the background which will always be there and does not require any protection.

For a small island, the United Kingdom has a wide diversity of habitats from highlands to heathlands, ancient woodlands to a long, varied coastline. We like to think of ourselves as a nation of green- fingered, pet owners who enjoy feeding the birds in the garden and visiting beautiful National Trust gardens on a weekend. However, in my lifetime, many people have lost their connection with the natural world preferring to get their natural fix from watching Springwatch or looking at a website or something on social media. Over this time the United Kingdom has become one of the most nature- depleted countries in the world.

The recent State of the Nature Report[1] reveals that between 1970 and 2013 56% of UK species have declined. When I was a child in the 1970s and 1980s, we lived opposite farmland. There used to be huge flocks of lapwing, oystercatchers, tree sparrows, and yellowhammers. Several hares lived in one of the fields and we would watch them chasing each other in the evenings. A hedgehog lived in our compost heap, and we regularly saw them at school and when we played on the common over the road. In the summer, our garden was alive with bees and butterflies. Now I get excited on a long journey if I see a lonely barn owl or solitary lapwing. Most times I do not even see a rabbit, something unheard of when I was in the back of the car on childhood road trips. Sightings of a stoat or badger are now so rare that I am thrilled when I see one. For my nine-year-old nephew's

---

1 The State of Nature Report involved 70 organisations. 7000 species were looked at. This report looked at England and Wales, but a similar report carried out in Scotland also showed that many species have seen dramatic declines.

The Wonders of the Wild Places

generation hedgehogs and turtle doves are something they only see in a book or on a screen. When my sister and I tell him about the huge flocks of goldfinches, greenfinches, and starlings we used to see regularly during our childhood it sounds like a fairy-tale. However according to Isabella Tree in her fantastic book, *Wilding,* I should not even compare my experiences to my nephew's generation but rather compare what I saw in my childhood with what my great- great grandparents would have seen during their childhoods. Isabella Tree summarises it perfectly when she says,

"We have an idea of how the countryside should look like and think it has always looked the same way, which of course it hasn't...We look at the countryside and see what is there, not what is missing. ...In our grandparents grandparents' generation every parish had species - rich meadows and coppiced woods. They regularly saw bitterns, corncrakes, turtle doves. They saw thousands of lapwing and hundreds of skylarks... We only have to go back four or five generations to see how much times have changed and how depleted of nature our landscape has become."

Remarkably though a recent survey showed a third of Brits have no awareness of the threats to our country's biodiversity. Another recent report found that that only 2% of people could identify common trees. Our great - great grandparents' generation would have known them all.

A shocking 1 in 7 species in the United Kingdom now face extinction – a quarter of mammals, a fifth of plants, and almost half of birds. In addition, 405 invertebrates and 232 fungi and lichen are in danger of extinction. The state of the Nature Report also found that 41% of species have decreased in abundance.[2] Through intensive farming, increasing industrialisation, population growth, deforestation, and lack of care and maintenance of our natural environment we face losing animals, plants, birds, fungi, and insects which we have shared this island with for hundreds, if not thousands, and, in some cases, millions of years. Animals, plants, birds, fungi, and insects which our ancestors could not only name but were essential to their way of life are in rapid decline and face extinction and many of us probably have not even noticed.

In the 1990s a fisheries scientist called Daniel Pauly coined the phrase,

---

2 It is not all bad news. 26% of species have actually increased during the same period.

'shifting baseline syndrome.' He realised that fishermen were comparing fish stocks with what they were at the beginning of their career, rather than the fish population in their original state. Pauly realised that each generation redefines what is 'normal' and forget that there used to be a lot more wildlife in the landscape.

With the United Kingdom recently leaving the European Union and losing the protections it gave to some of our most vulnerable habitats there is concern from many nature organisations that our natural environment will come under further threat from frackers, farmers, businesses, and developers.

This is not just bad news for the flora and fauna in this country, but also for the human inhabitants of this island as well. It is not a coincidence that as our natural world has become depleted the numbers of people suffering from diseases such as heart disease and high blood pressure, as well as depression and anxiety has increased. Recent research has shown that there are significant health benefits if we immerse ourselves in nature. When we are in a natural environment we release the bonding hormone oxytocin and beta-endorphins which lower our anxiety levels and raise our mood. This does not just keep us mentally fit. Professor Roger Ulrich at Chambers University in Sweden has studied the beneficial effects of nature on the body and discovered that just being in a green natural environment for a few minutes significantly lowered blood pressure and heart rates, and after half an hour the level of stress hormone, cortisol, significantly reduced too, which is not only good for our mental health but also for our immune system. The health benefits of connecting with nature are now so well known that doctors have begun socially prescribing so called 'green care' rather than conventional drugs for conditions such as depression.

Currently, the world is in the grip of a global pandemic which has not only led to millions of deaths worldwide but has also seriously damaged economies. The United Kingdom's economy is the worst it has been in over two hundred years. The government is promising a green economic recovery and vowing to combat climate change by projects such as the proposed new Northern Forest, whilst at the same time funding a project which is decimating ancient woodlands and nature reserves to build the controversial HS2 rail route which will allow people to travel from London to Birmingham slightly more quickly.

The Wonders of the Wild Places

I live on the edge of the beautiful North York Moors, home to one of the largest expanses of heather moorland in England and Wales which is internationally renowned as an important habitat for ground nesting birds. It is also home to the most northernly colony of the rare Duke of Burgundy butterfly and has the largest concentration of veteran trees in Northern England. However, despite the North York Moors being one of the fifteen national parks in the U.K. shockingly only 1% of the land is owned by the national park, with 80% of the land being in private ownership. In recent years, many of the inhabitants of this area have been campaigning against fracking within the area of the national park. Despite the government announcing a moratorium on fracking, people are still concerned about whether the precious habitats in this area are safe from development and industry.

As a country we value our built heritage and have various protections in place to maintain and preserve our old built infrastructure, but we have a very different attitude to our natural environment even though some of our veteran trees and hedges are much older than our castles, cathedrals, and stately homes, and our oldest tree, the Fothergill Yew[3], is probably older than Stonehenge. About a quarter of the land in England has some sort of legal protection which is meant to benefit the wildlife within it, with the government announcing their intention to increase this to a third of land by 2030. However, a lot of the land in these areas is currently in a poor state and suffering wildlife declines, due to issues such as intensive agriculture, overgrazing, pollution, and poor management. My worry is that once we have lost an important habitat not only do we lose biodiversity, but also an important piece of our history.

However, it is all not all doom and gloom. Things are being done internationally and nationally to restore and create habitats, and to reduce carbon emissions and pollution. Nature organisations, such as the Woodland Trust, Wildlife Trusts, and RSPB, are restoring and creating habitats, and there have been successful reintroductions of previously

---

3 In Perthshire. There is some dispute about the age of the yew because yews are very difficult to age but it is alleged that Pontius Pilate was born next to the tree.

rare or extinct species, such as red kites[4] and beavers. In some areas habitats are being left to restore naturally, such as at Knepp Estate in Sussex where farmland has been allowed to rewild, with fantastic results. Peatbogs and heathland are being restored, and trees are being planted. The government hopes to increase woodland cover over the next couple of decades, although there is criticism that there are not the resources to achieve their targets. [5] The government have also recently announced that post- Brexit, farmers will have to implement various schemes which protect and encourage nature in order to get government supplements.

We can still turn things around, but that means individuals as well as governments and organisations doing things which make a positive difference to the natural environment, such as having wild areas in our gardens or reducing our plastic usage. However, if this is to happen people must regain their connection with the natural world. A positive result of the covid-19 pandemic is that many people have had more time to spend engaging with nature. The aim of this book is to encourage people to get out and about and enjoy and appreciate the natural environment – you do not even have to go far- there are amazing things to see in your garden or local park. In the U.K. we share our island with approximately 70,000 species of animals, plants, birds, and fungi. After reading this book I hope you will go out and look at some of them.

In this book I describe common and rare species covering many different habitats. From the largest to the smallest species there are many surprises. From fungi to bees to trees, so many species are vitally important not only for biodiversity but also for our health and welfare, and even our survival. At the end of each chapter, I go into further detail about the threats to our natural environment but also provide information about things which you

---

4 There were only 5 breeding pairs of red kite in the UK at the turn of the 20th century; there are now 1424 pairs and numbers are continuing to grow, with red kites now seen in my birth county of North Yorkshire for the first time in many centuries.

5 The government has legislated to achieve net zero greenhouse gas emissions by 2050 and part of the way they want to achieve this is through planting millions of trees. If you are interested in how much has been achieved so far, the Woodland Trust article entitled Disappointing planting figures in England far below Government target, 11th June 2020 at www.woodlandtrust.org.uk is well worth a read.

The Wonders of the Wild Places

and your family and friends can do to combat these threats and help nature. You could join an organisation or build a bug hotel, plant a tree or some pollen -rich plants in your garden. There are resources at the back of this book to help you identify species. One of the things I enjoy about nature spotting is that you do not have to have specialist equipment, although a pair of binoculars and a camera or smartphone are helpful. When looking at nature please do not pick any flowers and remember to give the animal or bird plenty of space and try and stay as still and as quiet as you can. Some of the species described in this book can only be seen at night or at certain times of year. Some are so rare they can only be seen in limited places. However, wherever you live and whatever time of year you go out, there are amazing and interesting species out there to be seen. I hope you enjoy the wonders of the wild places.

If you go down to the woods today....

In the United Kingdom there are about 1.3 billion trees – 25 for every person – but it is still one of the least wooded countries in Europe with only 13%[6] of the United Kingdom being wooded, much less than the average of 36% across the rest of Europe. In England, most woodland is privately owned, whereas in Scotland only about a third of woodland is privately owned.

The first trees came over to Britain from mainland Europe about 10,000 years ago at the end of the Ice Age. Birch was the first to arrive, as it does now when new woodland is naturally created, followed by other well-known species – pine, hazel, elm, oak, and alder. The United Kingdom has the perfect climate and topography for tree growth, and trees grow faster here than the rest of Europe.

Opinions differ about how wooded Britain once was – some people think there was a mystical wildwood which covered most of the land, whilst others suggest that by about 2000 BC only about half the land was wooded. Woodland clearances probably started in the Neolithic period, about 6000 years ago, when people began to require land for agriculture. By the time the Doomsday Book was written in 1086 about 25% of Britain was wooded. However, following the Dissolution of the Monasteries in the 16th century, and increased industrialisation in the 18th and 19th centuries, during which woods were seen as a valuable commodity, many woods were felled, and by the First World War only 5% of the United Kingdom was wooded. The government felt that something needed to be done to rectify the situation, so the Forestry Commission was set up in 1919. It now manages 1.7million hectares of forest in Britain.

Only 2 % of the woodland in Britain is defined as ancient woodland. To qualify as an ancient wood there must have been a wood on that site since 1600 (or 1750 in Scotland.) Some of our ancient woodlands are very old. The Blean in Kent dates back thirteen centuries, and Hatfield Forest in Essex is about 2000 years old. The latter was a royal hunting forest,

---

6 10% in England, 15% in Wales and 19% in Scotland.

once owned by King Harold who was famously defeated by William the Conqueror at the Battle of Hastings. Another old royal hunting forest, the New Forest in Hampshire has a thousand ancient trees. This is thought to be the highest concentration of ancient trees in Western Europe.

Ancient woodlands are very important habitats, rich in biodiversity, which is why it is heart-breaking that HS2 will destroy 108 of these precious habitats.

Although most of the trees in these ancient woodlands will not be hundreds of years old some of them will. Ancient woodland is a good habitat for one of our trees with the longest lifespan- oak trees. There are two native oak species in the U.K. – English[7] and sessile oak trees. English oak tend to live in valley bottoms, parks, and woods, whereas the Sessile oak[8] is more of an upland species, mainly found in North and West Britain.[9] The acorns on an English oak have stalks, whereas sessile means 'sitting' because the sessile oak's acorn sits on the branch rather than on a twig. The leaves are also slightly different, with the English oak having a short stalk hidden by two small leaves at the base of the leaf, and the sessile oak having a slightly longer stalk.

Britain has more ancient oak trees than the whole of Western Europe; 118 in Britain compared to 97 in Western Europe. Sherwood Forest in the East Midlands has the largest concentration of ancient oaks in Northern Europe with nearly a thousand oak trees which are up to a thousand years old including the Major Oak, which is estimated to be between 800 and 1100 years old. This is said to be the biggest oak tree in Britain with a circumference of 36 feet (11 metres), and a canopy spread of 91 feet (28 metres). It is estimated to weigh a whopping 23 tonnes. [10] Not surprisingly, considering its location, it is said to have a connection with Robin Hood. According to legend, the famous outlaw and his band of Merry Men, and

---

7 Aka Pedunculate oak
8 Aka Durmast Oak
9 There are also the Turkey Oak, Holm Oak and Cork Oak in the U.K. All three species have been introduced to this country. See treeguideuk.co.uk for further details of these species.
10 Interestingly it is not called the Major Oak because of its size. It is named in honour of Major Heyman Rooke who wrote a book about oak trees in the area in 1790.

Maid Marion, used to camp underneath it. Whilst there is an ongoing debate about whether Robin Hood really existed, I am sure many outlaws and travellers have stopped a while to admire this incredibly special oak tree.

Mature oaks are extremely important because they support more lifeforms than any other tree. Five hundred species rely on an ancient oak tree. Other trees also provide valuable habitats for a wide range of species including one of my favourite trees, silver birch, which not only improves soil quality, making it easier for other plants to grow, but is also home to more than 300 species.

Oaks have always been seen as the King of trees and many cultures have associated them with thunder, probably because oak trees are more prone to being struck by lightning than any other tree, possibly because they quite often are found in the middle of fields, or maybe because they tend to be one of the tallest trees in the forest.

A mature oak tree is estimated to have 700,000 leaves. However, it can expect to lose half these leaves a year because they are an important food source for caterpillars. Unfortunately, in the case of some of these caterpillars, this can be damaging for the tree, and for humans. Currently, in the London area and a handful of other areas of mainly Southern England, the caterpillars of the oak processionary moth are causing problems. The caterpillars can strip the tree, making it vulnerable to other pests and to drought. The caterpillars can also cause rashes and eye irritation to humans and animals. [11] However, oaks and other trees, have an arsenal of ways to repel pests. Some trees, like holly, have prickly leaves at the bottom of the tree to make it difficult to be nibbled; other trees, including oaks, have tannins. Tannins bind with proteins to affect the animals' feeding behaviour. However, rodents and rabbits have evolved so that they have an ammino acid called proline which binds with the tannin but blocks it from affecting them. Tannin, and other chemicals including limonoids in citrus trees, also acts as natural insecticides. Even cleverer, trees can identify insects by their saliva. They then release pheromones to attract the predator of the pest which is trying to eat them. However, extreme weather conditions, such as drought or flooding, can affect the

---

11 For more information look at www.forestresearch.gov.uk/oak processionary moth.

tree's ability to effectively carry out these various techniques of repelling pests, so as we continue to experience more extreme weather conditions because of climate change, we should expect to see an increase in pests. At the end of the chapter, I will further discuss some more of the threats which trees are facing.

If oaks are regarded as the Kings of the forest, beech trees are thought of as the Queen of trees. They are in the same family as oak trees, and like oaks are long livers. Scientists have estimated that in the lifetime of a 400-year-old beech it will produce 1.8 million beech nuts, however, incredibly, only one of these will grow into a new beech tree. This is because beech nuts are such a valuable source of food to so many animals and birds, and even to humans who can grind the nuts to make flour. In recent years scientists have discovered that oak and beech trees can communicate over huge distances. They agree to increase or reduce the number of acorns or nuts they produce each year. Thus, every five years they produce a bumper crop. These years are called mast years. In doing this they help to regulate the populations of the animals which rely on the nuts and acorns for food.

So, how do the trees communicate, and is it just restricted to oak and beech trees, or can other trees communicate too? The answers to these questions are not straightforward. Research is still ongoing, and opinion differs, but it seems that deciduous trees[12] in a natural woodland, as opposed to ones which have been planted in a plantation, can communicate in various ways, including soundwaves and electrical signals, which travel slowly at a third of an inch a second, along what has been described as the 'wood wide web'.[13] This describes extensive networks of fungi underneath the forest floors, linking up various trees. In his wonderful book about fungi, *Entangled Life*, Merlin Sheldrake expresses the opinion that the term 'wood wide web' is misleading because it suggests that the plants are like routers, and the fungi acts like cables. This implies that the fungi are just passive parts of the network, but this is far from the truth. As Merlin Sheldrake explains, "Every link in the wood wide web is a fungus with a life of its own." In reality, 90% of all plants, including trees, have

---

12 Trees which lose their leaves in autumn
13 The term was first coined in 1984 by an author about mycorrhizal biology, David Read.

symbiotic relationships with fungi. These are known as mycorrhizal fungi which comes from the Greek words for fungi and roots.[14]

What we think of as fungi is actually their fruiting body, either a mushroom or toadstool. Underneath the ground the fungi have fine thread-like structures called hyphae. These are five times thinner than a human hair; only between 2 and 20 mm in diameter. These hyphae entwine to form networks known as mycelium. No two mycelium networks are the same. Some could fit into a speck of house dust whereas others cover huge areas. In America, the honey fungus has been found to cover 2200 acres. The mycelium tends to spread out in all directions from a central point. Normally the only thing that can be seen on the surface are the fruiting bodies, which can sometime form fairy rings of toadstools.

These networks link up various plants, including trees. However, one plant or tree might have several such mycorrhizal networks tapping into their roots. According to Peter Wohlleben in his fascinating book, *The Hidden Life of Trees,* in an oak forest each tree could have as many as 100,000 different fungi living amongst its roots. Plants and fungi both rely on their senses. The plant's shoots and leaves explore the air, obtaining information which they can then use to adjust their behaviour, and their roots spread out underground looking for connections with different species of fungi. Meanwhile the fungi's hyphae spread out in different directions growing, thickening, changing direction, and even pruning hyphae if they are not being used. In one way this can be seen as being a bit like a brain. These symbiotic relationships between plants and trees are beneficial to both parties. The trees (and other plants) are exposed to light and air so ,at a basic level the plant (a tree in this case) is able to convert light and carbon dioxide into sugar and other carbohydrates, which it gives to the fungi; in return, the fungi breaks down rocks and decomposing material and provides the tree with valuable minerals. One of the chemicals which the fungi give to the host plant is phosphorus which not only allows the plant to grow more quickly, but also helps the plant to extract more carbon dioxide from the atmosphere, something which is vital in combating climate change. The fungi also provide the

---

14 *Mykes* for fungi and *rhize* meaning roots.

plant with up to 80% of the nitrogen it needs, as well as zinc, copper, and water. In return the plant provides the fungi with about a third of its carbon. In a hectare of woodland 280kg (0.28 tonnes) of carbon can be transferred through these mycorrhizal networks. This is the same amount of carbon used to power an average home in a week. However, like many things in nature, it is not straightforward. Whilst fungi help to transfer and store carbon, it is also a major source of carbon emissions. As fungi break down dead wood, they emit 85 gigatonnes of carbon into the atmosphere every year. In comparison in 2018 the global combustion of fossil fuels emitted 10 gigatonnes of carbon.

These relationships are not just reduced to an exchange of food. Many other things also travel along these complex networks. Infochemicals, which convey information about the plant's condition which can then be used to fight predators, can be passed along mycorrhizal networks; as can glutamate and glycine, which are the most common neurotransmitters in human brains and spinal cords, but also major signalling molecules in plants. In *The Hidden Life of Trees,* Peter Wohlleben suggests that trees use the wood wide web to communicate with other members of their family, providing sick trees with extra nutrients. Merlin Sheldrake in *Entangled Life* does not like the humanisation of plants, but acknowledges that it seems that communication is happening, although maybe not as humans would understand it.

There is still a lot of research being done in this area. Many experiments so far have just taken place between plants in a lab and Lucy Gilbert and David Johnson at Aberdeen University think more evidence is needed before it can be shown that forests can communicate via wood wide webs. One of the few people who have studied it on a larger scale is Kevin Bieler, who has studied a Douglas fir forest in British Columbia in Canada. He discovered that there was a connected wood wide web, but not all of the trees had the same number of connections. Older trees had more connections than younger trees, making them important hubs in the network. If an old tree is cut down it affects the rest of the network. Scientists have also shown that related trees have more connections than unrelated trees.

The study of fungi, known as mycology, is in its infancy and I suspect that we will discover much more about the wood wide web in the next few years. One of the issues when referring to fungi is that it is extremely difficult to define. Until the 1960s fungi were classified as plants, but science has shown this is not accurate. However, because fungi can grow in a variety of forms it is difficult to pin down exactly what constitutes fungus. This probably explains why it is thought that about 90% of fungi have still not been documented.

Fungi is not just incredibly fascinating underneath the ground, their fruiting bodies, are also interesting. These fruiting bodies are mushrooms or toadstools. The difference between the two is etymological rather than scientific. Toadstools tend to be the word used to describe toxic inedible fungi, whereas mushrooms tend to describe the edible fruiting bodies.

One of the toadstools that most people are most familiar with is the classic 'fairy' toadstool with its red spotty top. Its real name is red fly agaric. They particularly love birch woodland because they have a symbiotic mycorrhizal relationship with birch trees. It is poisonous so do not touch it. It was used as an insecticide. Victorians and Edwardians loved red fly agaric and it often featured on Christmas cards because it was a symbol of good luck. It is alleged that its colour scheme inspired Santa's famous red suit.

What you might not know is that this toadstool, which you can commonly spot in woods in the autumn, can break down some of the most durable substances in nature – cellulose in plants, animal bones, the hard shells of insects and even rock. This is one of the most amazing things about fungi. As Merlin Sheldrake says in his book *Entangled Life*, "Fungi are metabolic wizards...." By using enzymes and acids fungi can break down a wide range of substances including crude oil, polyurethane plastics, and even T.N.T. There is even a species of fungi which is resistant to radioactive matter and is being looked at as a way of cleaning up radioactive sites, such as Chernobyl. Scientists all over the world are looking at ways of using fungi from cleaning up oil spills, dealing with the plastic waste problem, and filtering water. White rot fungi are also being studied to see if it can be used to help stop bee mortality because it its antiviral properties. There is also a company called Evocative which is using fungi to decompose waste which would otherwise have gone to landfill, to grow furniture and packaging. Two companies, Biohm and

Carbios, are also developing ways of using fungi to break down P.E.T.-plastic waste. In short fungi are brilliant organisms and I think the next decade might see them being used even more; using nature to deal with the problems humans have created.

Another mushroom that you might see in British woodlands is the gloriously named stinkhorn mushroom. These phallic looking mushrooms greatly upset the sensibilities of Victorian ladies. Charles Darwin's daughter, Etty, was obsessed with them, walking around the lanes and woods near where she lived, removing any of the offending mushrooms which she saw and burning them on the fire. However, these mushrooms have an incredible superpower. They can grow through asphalt roads. In order to do that they need to exert a force which would enable them to lift a 20 stone (130kg) object (the equivalent to a human lifting half a pig).

In pine forests you might see the beautiful yellow staghorn, normally on a tree stump. In broadleaf woodlands look out for the candlesnuff fungus, which as the name suggests looks like a snuffed-out candlewick. This fungus has antiviral properties and is currently being used to treat some human cancers. Also look for beefsteak fungus, which looks like raw meat, or chicken-of-the-woods which looks like the feathery yellow bottom of a hen. After rain you might see for the common inkcap which grows on dead and decaying wood. As the name suggests it was used to make ink for important documents because the spores can be seen under a microscope, thereby making forgery much harder. Other common fungi, which is easy to spot, are the self -explanatory jelly ears which love beech and elder woodland; dead man's fingers, mistakenly thought to be someone trying to escape from their tomb, which can be seen in Spring and Summer in deadwood in beech woodland, and King Alfred's cakes which are black fungi found on deadwood mainly in beech and ash woodland.

You do have to be careful when looking at toadstools and mushrooms because we do have some deadly ones in the U.K. The well-named death cap is the deadliest known fungi. It grows across the U.K. in broadleaved woodlands and ingesting even just a little bit of it is fatal. To be fair the names of the most of the deadly toadstools normally give you a clue that they should not be ingested – destroying angel, funeral bell, deadly webcap, and angel's wings. However, the problem is that a lot of the deadly fungi look remarkably like non-toxic fungi and can often be mistaken. The lesson to be learnt from this is not to pick and eat fungi. If you are interested in learning more go out with an experienced forager.

Another species of tree which is often found in ancient woodland are hazel trees. Since the Bronze Age this is one of the species of tree which has been used in coppicing. Coppicing is a practise where the lower outer branches of the tree are harvested on a cyclical rotational basis and the wood is used for firewood and to make furniture.[15] The heyday of coppicing was the 16th and 17th century when coppiced woods were used to provide charcoal for the iron industry. Coppicing provides a sustainable useful product but is also good for the individual tree, allowing it to live longer, and for the woodland in general because it leads to an increase in biodiversity. Since the Second World War we have lost 90% of our coppiced woodland in the U.K. but, fortunately, in recent years, as nature organisations look at ways of increasing biodiversity, there is a move back to coppicing woods. I live opposite a deciduous coppiced wood which is home to a wide range of species.

Another species which can be a good indicator of an ancient biodiverse woodland are one of my favourite flowers, bluebells. The United Kingdom is one of the best places in the world to see these beautiful, sweet-smelling flowers which carpet our ancient woodlands in mid-April to mid-June. Their Latin name is *Hyacinthoiden non-scripta*. Legend has it that the Greek God of Sun, Music, and Poetry, Apollo, fell in love with a young man called Hyacinth. As they were throwing a discus to one another, Zephyrus, the jealous God of the West Wind, caused the discus to crash into Hyacinth's head, killing him. Apollo was devastated. He immortalised Hyacinth by turning him into a blue flower. The Greek legend says that as Hyacinth's blood drained into the soil a blue flower sprang up, which Apollo inscribed with his grief, writing AI AI (alas) on the petals. British bluebells have no such inscription.

Since the Bronze Age glue made from the bluebell's sap was used to attach feathers onto arrows. The sap was also used for centuries to bind

---

15 This is different from pollarding where the top of the tree is removed.

books, and in Elizabethan times it was used to starch the large ruffs (collars) worn by higher levels of society. In Victorian times there was a craze to attribute various characteristics to different flowers. Therefore, bluebells were associated with everlasting love, but they were also associated with fairies, and it was said that if you heard the dainty bluebells ring, your death was imminent. Today we might associate the rose with St George's Day, but actually it is the bluebell, which is associated with the saint, maybe because in good years they tend to be out at the end of April, when it is St George's Day.

Unfortunately, our bluebells are under threat from Spanish bluebells, which are paler and have different shaped flowers. These pesky invaders are cross fertilizing with British bluebells, creating a hybrid. They are in the wood over the road from where I live and the only way of removing them is to dig them out, making sure you take all the roots too. However, make sure you are removing the right ones. English bluebells are protected by law.

Where you see bluebells, you often see another flower synonymous with ancient woodlands in springtime, ramsons, or wild garlic. These pungent white flowers are an excellent food source for insects and humans. All parts of the plant can be eaten in salads and the bulb has been shown to relieve rheumatic pain and lower cholesterol.

Two other species which are easily seen in British woodlands in spring are wood anemones, and wood sorrel. Another name for wood anemones is windflower because there is a Greek legend that Zephyr, the West wind, fell in love with a nymph called Anemone. The goddess Chloris turned Anemone into a flower in a fit of jealousy, and left her to the mercies of Boreas, the North wind. The Romans had their own legend about the flower. According to them, Adonis was mortally wounded by a boar. His lover, Venus, was so distraught and vowed she would turn him into a flower so that he could live forever. The flowers were created by her tears. Subsequently wood anemones are associated with love. Wood anemones are a good way of forecasting the weather because they tend to close if rain is one the way. Their roots spread quite slowly, about six feet every century, so you can gauge the age of the wood by how far they have spread across the woodland floor. The leaves of the beautiful white-bell flowered wood sorrel are edible and taste of apples.[16] According the 17th century botanist and herbalist, Nicholas Culpepper, wood sorrel can be used to reduce inflammation, stop bleeding and vomiting, and aids digestion. This is one of the reasons why it is so important that we protect and value our natural habitat. There are so many plants which have medicinal properties, and modern-day scientists are only beginning to understand how useful many of these plants are for treating severe illnesses, such as cancer and diabetes.

Snowdrops are one of the earliest sources of pollen and can be seen carpeting woodlands in February and sometimes even earlier. The reason they can survive in low temperatures when so many other plants cannot is that they have in-built anti-freeze which protects the plant from being damaged by ice and frost. Their Latin name is *Galanthus* which means milk-flower of the snow. Snowdrops seem such a part of our late winter and early spring woodlands that many people probably think they have always been there, but, in fact, they were introduced to the Britain in the late 15th century, probably by Italian monks, and the first wild snowdrops were recorded as late as the late 18th century. As well as brightening up stark woodland these delicate flowers might also be of huge medical significance. Scientists in Eastern Europe have extracted a substance called galantamine from snowdrops. Galantamine can be used to stimulate the regeneration of nerve cells and scientists hope that it can be used to treat many neuralgic and degenerative disorders, such as Alzheimer's. It is amazing to think that the ubiquitous snowdrop might be responsible for a major breakthrough in medical science.

Another plant found in woodland in herb Robert. There is some contention as to which Robert this plant commemorates, with some people claiming it is Robert, Duke of Normandy, who allegedly used the herb to cure the plague; others claim it is named after a French Saint, Saint Robert of Molesme whose Saint's Day is the 29th April, which is when this plant normally comes into flower. Saint Robert apparently used the herb to treat lots of different conditions, and it is a useful medicinal herb, from

---

16 Always take care when eating wild things. It is best to go out with a professional forager just to make sure you get the right species.

staunching blood to treating kidney and bladder infections to treating inflammation.

Violets are a common plant in British woodlands and can also be seen in hedgebacks and grass verges. Indeed, their name derives from *vias* meaning wayside. Since Ancient times they have been used for perfume and in Ancient Greece a garland of violets and roses would be placed on the door of a house where a new- born boy lived. Like several other plants and trees violets contain salicylic acid (which we know as aspirin) and they were also used to treat inflammation and mouth ulcers, as well as bronchial problems and insomnia. The leaves are an antiseptic.

As well as the biodiverse ancient woodlands in Britain we also have two very rare types of woodland which provide habitats for some of our rarer species. On the Western side of Britain, we have Atlantic rainforests which are remnants of the last Ice Age and are rarer than tropical rainforests such as the Amazon. On average these rainforests get 11 ft (3.5m) of rain a year; much more than the national average of about 4ft (122 cm). Many have been lost because of overgrazing. However, in the last forty years the Forestry Commission has begun to slowly restore these important habitats by thinning rhododendrons and conifers and planting native trees. These changes must be made slowly in order not to affect the humidity of these precious environments. These beautiful rainforests are home to five hundred species found nowhere else in the world. High rainfall and mild temperatures create a humid environment which is the perfect conditions for fungi, lichens, and liverworts, which in turn regulate the temperature of the Atlantic rainforest.

Lichens are one of the oldest organisms in the world and grow in even the harshest of environments. When a volcanic eruptions causes lava to form new land mass it is lichens which are often one of the first things to grow there, making the land inhabitable for other species to follow. As Merlin Sheldrake eloquently puts it, "Lichens are the go-betweens that inhabit the boundary between life and non-life."

If all the lichens in the world were spread out, they would cover 8% of the world's surface, an area larger than the world's tropical rainforest. However, like fungi, lichen is very difficult to define. In the 19th century a Swiss botanist called Simon Schwendener, believed that lichen consisted of two separate organisms- fungi and algae- which had a symbiotic relationship which helped them both to thrive in conditions where they could not exist on their own. The fungi provided the physical protection and the nutrients, whereas the algae harvested light and turned it into carbon dioxide which it made into sugars for energy. However, Schwendener did not believe it was an equal partnership, describing the fungal part (mycobiont) as, "parasites with the wisdom of statesmen" and the algal partners (photobiont) as "slaves." This theory caused controversy at the time because it seemed to go against Darwin's Theory of Evolution which suggested that species diverged from one another, whereas Schwendener suggested that the different organisms were converging to form a new species. Many people, including the knowledgeable mycologist, author, and illustrator Beatrix Potter, were dismissive of Schwendener's theory, but it remained the dominant theory for about a century. However, in 2016 Toby Spribille, an American lichenologist, discovered that they are actually far more complex than Schwendener thought. Spribille discovered that lichens actually have several symbiotic partners. As Merlin Sheldrake puts it in his excellent book *Entangled Life*, "...lichens confuse our concept of identity...," and can be thought of as, "micro-planets." There is not a single definition of what lichens are because the partnerships which form lichens vary with some being a partnership between fungi and yeast, others partnering with algae, and bacteria also joining in the partnership. Every lichen is different, forming symbiotic relationships with sometimes multiple partners in order to survive. One in five fungi lichenise, although even this is not straightforward. Some species, like penicillin, used to lichenise but no longer do, others can switch back and forth, and others can live as lichens depending on the circumstances.

As well as seemingly going against Darwin's evolutionary theories, and being pioneers and ecosystem creators, lichens are also amazing organisms (or more accurately, a partnership of organisms) in other ways too. They have layers of tissue which can block out cosmic rays. They also produce a thousand chemicals which can be found no- where else on earth. The lichens use some of these chemicals as a sunscreen. Other chemicals are used by humans in perfume, dye, food, and antibiotics. However, for me, the most amazing thing about these incredible natural anomalies is that they can be resuscitated after being dehydrated for a decade. Lichens are survivors. They have existed for at least 400million years and whatever happens with our natural environment over the next few decades, it seems a good bet that lichens will still exist, maybe even in a different form again, when many other species have become extinct.

Lichens love ancient woodlands, and Atlantic rainforests, because there are hundreds of species in Britain which really like alkaline conditions. Tree bark becomes more alkaline over time and ash has a particularly alkaline bark. Over five hundred species of lichens just live on old ash trees. Lichens also like old woodland and undisturbed environments because they take a long time to develop, only growing up to 2mm per year.

Lichens are very sensitive to pollution, although they overcome it by becoming crustier; however, some finer lichens can be a sign of better air quality. A grey-green bushy lichen, called old man's beard, is often seen draped over the branches of trees. It also likes younger trees because it is an early coloniser. It is a sign of purer air quality. Common orange lichen can often be seen on trees, especially elders, as well as one roof tiles, wooden fences and on rocks. Another well- known grey-green lichen, known as monk's hood lichen, can often be seen in woodlands. It prefers trees with acidic bark, and it is very sensitive to nitrogen pollution so, again, if you see it, it is a sign of good air quality.

Also look out for mosses and liverworts when in Atlantic rainforests. They are both members of the bryophytes family, which are the only living things related to the first plants, which about 500 million years ago only lived in the water and, with the help of algae, progressed to land. Until recently liverworts and mosses were thought to be closely related, but recent D.N.A. evidence has surprised biologists by revealing that liverworts were actually the first plants to evolve on land, with most of the plants in the world today evolving from them; whereas mosses have characteristics which are not shared with liverworts, suggesting they might have evolved differently. Despite this they do have characteristics in common. Unlike other plants they do not have roots. However, mosses have filaments called rhizoids which enable them to anchor onto trees and rocks; liverworts also have rhizoids although theirs are only a single cell. Mosses' leaves grow in a spiral whereas liverworts have lobed leaves with side of a stalk.

The Wonders of the Wild Places

Mosses and liverworts provide valuable habitats for many other creatures, insects, plants. They act as sponges, capturing moisture, and then slowly hydrating the tree. This can be particularly useful to the tree during periods of drought. Like lichens, mosses are hardy, early colonisers, and survivors. They can withstand long periods without water then burst back into life. They have even been known to rehydrate after fifty years without water.

Scotland's Caledonian forests are another extremely rare and biodiverse habitat. These are Scot's pine forests which are descended from some of the first trees to have arrived in Scotland in about 7000BC. The habitat remained largely unchanged until the 17th century, creating a unique ecosystem which is home to many of the rarest species in Britain.

Scot's pines are the U.K.'s only native pine tree and can live up to 700 years. They are monoecious, which means that the male and female flowers grow on the same tree.

These are commonly seen in my home county of North Yorkshire, in forests and along old drovers' roads.

The Caledonian forests are home to the only bird endemic to the United Kingdom, the Scottish crossbill. These were only recently found to differ from the common crossbill, which lives in the Northern Hemisphere. The Scottish version has a larger beak and makes a different sound to its more common cousin. As the name suggests Scottish crossbills have crossed mandibles which allow them to eat seeds from pinecones. Like a lot of rare species, they can be difficult to spot. Not only can they only be seen in the Scottish Highlands, but they also tend to hang around the top of the trees so are difficult to see without binoculars. Whatever expensive equipment, when trying to look for wildlife you have it all depends on being at the right place at the right time which is why I was delighted to see a Scottish crossbill a few years ago as I walked through a Caledonian forest. It rather conveniently flew down to a lower branch and chirped at us, making it a lot easier to spot.

Another rare species which was thought just to live in these forests is the pine marten. They were believed to be extinct in England, until recently being reintroduced to the Forest of Dean. However, there has been a colony of them near where I grew up in North Yorkshire, for at least 35 years. Seven years ago, as I walked through a mixed broadleaf and Scot's pine woodland, just a few miles from where I live, I was fortunate to see one of these beautiful mammals in broad daylight. I can honestly say it was the best animal encounters of my life. From being a child, I had a list of animals I really wanted to see and pine marten was top of the list. On many childhood holidays in Scotland, I would peer at the top of trees, hoping to see one. One of the things that made this difficult was that they are normally nocturnal, which made it all surprising to see one at ground level on a bright autumnal afternoon.

Pine martens are part of the Mustelidae family, which also includes stoats, badgers, and weasels. They are distinctive because of their white bib, bushy tail, and pert bushy ears. These rare animals have been in Britain since the last Ice Age. They prefer conifer woodland and have flexible joints which can turn 180 degrees, allowing them to climb trees. They have a thick fur which helps them stay warm in Winter because they do not hibernate. Dr Joshua Twining at Queen's University, Belfast has discovered that in harsh winters they can delay an embryo until the worst of the weather has passed, allowing their young a better chance of survival. Pine martens have large territories, with a male pine marten covering 15.5 miles (25km.)

In 2015 pine martens were reintroduced into Wales. Pine martens are also colonising parts of Northern England and now they have been reintroduced into the Forest of Dean in Gloucestershire, so hopefully, with time, they will become a more common sight. They may also play an important part in the long-term viability of another species which was reasonably common in the North of England during my childhood in the 1980s, but whose numbers have declined dramatically in the last few decades – the iconic, and very cute, red squirrel.

Red squirrels are always thought of as native to Britain, having lived here for about 10,000 years, however, on a recent episode of the TV show, *QI*, it was revealed that many of the red squirrels in the U.K. today are descended from Scandinavian red squirrels introduced about a century ago because of another dramatic decline in their populations.

As well as habitat loss the main threat to these popular mammals are the non-native grey squirrels, introduced to Britain in the 1870s. The red squirrel's larger American cousins, grey squirrels, are much more robust, being able to eat nuts with a much higher tannin content which provides them with more protein and energy. However, the main threat which the greys pose to their smaller red cousins, is the parapoxvirus, also known as squirrel pox. This disease was first noticed in Norfolk in the early 1970s and has unfortunately spread across Britain. The problem is that grey squirrels are carriers of this awful disease, which produces scabs and sores around the squirrel's eyes, mouth, ears, feet, and genitalia, eventually making it difficult for them to see or feed properly, but grey squirrels are immune to the effects of the disease. It is still not fully understood why the grey have immunity to squirrel pox, it might be that they have been exposed to it for a longer period so have developed antibodies which make them immune to it. What is clear that unless action is taken red squirrels could become extinct in the U.K. within a decade.

Over recent decades conservation charities have tried culling greys, but this has been expensive and has not really made any material difference. Scientists are trying to use the antibodies found in grey squirrels to develop a vaccine but even if they achieve in doing so it is not known how reds will actually be vaccinated. Recently scientists have looked at other ways of controlling the grey squirrel population. One way, just approved by the government, is putting contraceptives, disguised in hazelnut spread, in woodlands. It is hoped that this will not only help red squirrels but also British woodland. Grey squirrels have a habit of stripping bark from young trees, seriously damaging, and in some cases, killing the younger trees, thereby damaging valuable native tree woodland habitats. The contraceptive laced spread is placed in special feeding stations, accessible only to grey squirrels. This project is yet to be rolled out but, if successful, it is hoped that it will reduce the grey population by 90%, giving a significant boost to the red squirrel populations, which are now mainly confined to Cumbria and Northumberland, Scotland, and the Isle of Wight.

However, another way of controlling grey squirrel populations might be the mass introduction of pine martens because they eat grey squirrels. Dr Twining and his team at Queen's University in Belfast have discovered that pine martens target greys exclusively during the spring and summer when it is the squirrels' breeding season. Although they have been known to eat red squirrels too, they eat the reds in much smaller numbers because the red squirrels are smaller and faster. Again, this project is just in its infancy, but it will be interesting to see over the next decade or so whether it makes a big difference to red squirrel populations.

It would be sad if we could no longer see red squirrels in Britain. As well as being cute mammals, with their tufty ears and bushy red tails, they are also interesting animals. They have doubled jointed ankles which enable them to run down trees headfirst. They are also strong swimmers. Recent research has shown that, like humans, squirrels can be left or right- handed. However, researchers at Exeter University discovered that squirrels which strongly lateralised (preferred one side over the other) were less good at performing learning tasks, compared to the ambidextrous squirrels.

Incidentally there is a third type of squirrel which can be seen, mainly in Eastern England (although I saw one in Derbyshire), the black squirrel. The first one was recorded about ninety years ago. It is thought to be caused by a genetic mutation in a grey squirrel. It seems that female grey squirrels are attractive to black squirrels, which are high in testosterone, meaning that the mutation might be passed on, and the black squirrel population might increase. Unfortunately, this is not good news for the red squirrels because black squirrels also carry the parapoxvirus.

The rarest mammal in Britain lives in the Caledonian forests of Northern Scotland, although if something is not done soon they will become extinct in the near future. The Scottish wildcat is thought to be reduced to under a hundred individuals, making them rarer than Sumatran tigers, and making their population unsustainable. A pure wildcat has a thicker tail with thin black stripes and a black tip. I say pure because one of the problems is these wildcats, which have been in Britain since the last Ice Age, have interbred with feral domesticated moggies and it is now thought that most of the population are hybrids. This not only dilutes the gene pool, but also means that the wildcats are unable to fight the diseases which domesticated cats spread. However, once again, all is not lost. The Scottish Wildcat Conservation Action Group has a plan. They have bred wildcats with the purest native stock available and hope to interbreed them with wildcats which have been imported from the Continent. These are genetically different to Scottish wildcats, leading to criticism about whether the Scottish wildcat can be saved as a distinctive species. It is hoped that these captive wildcats will be released into the wild in 2023 interbreeding and strengthening the existing Scottish wildcat population. Before this can be done feral domesticated cats in areas known to have existing wildcat populations are being vaccinated and neutered. The Wildwood Trust also hopes to release captive – bred wildcats into England and Wales in the next few years, after them being extinct in England for 200 years.

Another rare mammal which can be found in woodland are hazel dormice. These tiny mammals, less than 3 inches long, typically spend seven months a year asleep, slowing down their heartbeat so it exists in a torpor during this period. Their populations have declined by a huge 60% this century, mainly due to loss of habitat. They currently are just in small pockets of the South of England, but there is a breeding project taking place in the Blean ancient forest in Kent. Hazel dormice have already been released in the Midlands and are due to be released in Cumbria. So far, the reintroductions are being very successful so hopefully within the next decade more of us might have a chance of seeing one of these tiny creatures.

We do not have many large land predators in Britain. The largest land predator are badgers, although their main diet are earthworms; they can eat hundreds per night. Sadly, during the last decade these mammals have been unfairly accused of spreading T.B. to cattle and thousands have been culled. This has had no effect on T.B. levels in cattle and has cost a lot of taxpayers' money. Recently the government have announced a change of policy which will hopefully allow these beautiful nocturnal animals to live alongside livestock. This would be good news for our natural environment because recent research has shown that badgers improve biodiversity and soil health. In creating the setts where they live, they create spoil heaps which retain moisture and are important habitats for amphibians, invertebrates, and pollinators as well as redistributing seeds. The inside of the sett also provides a home for other mammals, plants, and fungi.

Of course, it is not only animals and plants which live in our woodlands. They are also home to many of our birds. Some of these birds are annual migrants to the United Kingdom. The chiffchaff, named after the call it makes, is one of the heralders of Spring. Despite weighing less than 0. 3 ounces (8g) it flies from Africa to our broadleaf woodlands every Spring.

Another visitor from Africa, synonymous with Springtime are cuckoos. These birds have decreased in recent years because they cheekily lay their eggs in other birds' nests, specifically meadow pipits, dunnocks, and reed warblers, and these birds have declined by 50% in my lifetime. Another problem is climate change which affects not only the start of the season, but when the host birds are nesting. The surrogate parent birds are now nesting before the cuckoo returns to Britain. As if this is not bad enough the hairy caterpillars of the garden tiger moth which they eat have also declined by a whopping 92% in the last sixty years. I feel extremely fortunate to be able to hear a cuckoo every Spring. There are at least two in the town where I live and one sits on a tree near my house making its familiar call, hoping for a mate. I wonder though for how much longer I will be able to hear this bird which has had such a huge influence on our culture, inspiring, amongst others, Shakespeare, and Wordsworth, and even entering our language either to mean someone who is being unfaithful or someone who is stupid[17]. Our literal associations with cuckoos go back a very long way. The cuckoo is one of the few birds to be referred to in Anglo-Saxon literature and appears in the oldest known written verse in English, appropriately called the Cuckoo Song, which was written in 1225, only a decade after the Magna Carta was written. It also appears in one of the first English dictionaries written in 725 AD. Back then it was known as *Gaec*, an Old English name derived from the Norse name. The cuckoo is still known as *gowk* in Northern England and Scotland.

A bird particularly associated with ancient oak woodland is one of our most colourful birds, the jay. These members of the corvid family have the Latin name *Garrulus glandanus* which is translated as the chattering acorn gather. This is a good name for this clever bird which not only mimic other birds, and have conversations between themselves, but also play a key part in forming new woodland. Jays can carry six acorns at a time.

---

17 The term cuckoo to refer to someone who is stupid or crazy dates back to at least 1580.

The Wonders of the Wild Places

They bury the acorns about 1.5-3 feet (46-91cm) underground, too deep for most acorn-eaters, about 60-70 yards (55-64m) from the tree. They rely on these stores of acorns during the Autumn and Winter, but the rest of the year they are less reliant on these buried acorns, which germinate allowing many new oak saplings to grow. It has been calculated that a single jay can distribute more than 7500 acorns in a month. Although not all of these will become a sapling it still plays a significant role in allowing the regeneration of our native broadleaf woodlands. As the saplings grow the jays nibble the primal leaves. This would kill most saplings, but unlike other trees oaks do not need these primal leaves to enable them to grow. In a recent experiment to rewild an arable field to, jays were responsible for planting about half the trees in the new woodland.

Two of the jays cousins, jackdaws, and magpies, can also be seen in woodland. These birds are good examples of a trend which was popular in medieval times to humanise birds and animals by giving them a human name. Thus, wrens were called Jenny, sparrows were Philip (probably because of the sound it makes), goldfinches were called King Harry's Red Cap (because of their plumage), and green woodpeckers were known as Laughing Betsy. Prior to this jackdaws had just been called daws, probably because of their call. Jack was often used to describe a small or male bird or animal, whereas Jenny was used to describe a female bird or animal. Magpies were called *pyge* which became pie, probably from the Latin *pica* meaning bird. From the mid- 16th century pies became Margaret Pie which was shortened to Meg or Mag Pie. There is a link with the term of a talkative woman, a chattermag, probably because magpies make a chattery call. Interestingly, by the 17th century jackdaw was used to describe the male version of chattermag, a talkative and obnoxious person.

Like cuckoos, magpies have become ingrained in our literature and culture, and are especially associated with luck. In many areas it is tradition to salute a loan magpie and to say good morning to it. Many of us will be familiar with the rhyme which starts, "one for sorrow, two for joy..." This rhyme first appeared in print in the late 18th century but is probably older in origin. There are many regional variations to the rhyme. The original rhyme, written by a vicar and antiquarian called John Brand, was just four lines long, "One for sorrow, two for mirth, three for a funeral, four for a birth." This was added to in 1846 by Michael Aisable Denham, a collector of folklore, who included, "...five for heaven, six for hell, seven for the devil, his very own self."[18] The version I learnt, growing up in North Yorkshire is, *one for sorrow, two for joy, three for a girl, four for a boy, five for silver, six for gold, and seven for a secret which has never been told.* However, there are numerous other regional versions although there are similarities, the rhyme oscillates between good and bad luck depending on the number of magpies seen, quite often referring to birth and death, heaven and hell, rich and poor, and England and France. The reason for this might go back a very long way. Greeks and Romans once associated magpies with good luck, whereas early Christians alleged that magpies were the only birds who refused to comfort Christ on the cross, unlike robins who got so close that his blood stained their breasts, and associated magpies with bad luck. In medieval times magpies were often seen scavenging near battlefields and gallows which also led to an association with death.

Another bird which can be heard more often than it is seen in woodlands are tawny owls. The familiar *twit twit twhoo* call is actually two owls, with the female tawny providing the *twit* and the male replying with a *twhoo*. The call informs the owl much more than just the gender of the other owl though, tawny owls have a range of hoots which can tell them about the weight of the bird and even how aggressive it. Scientists have discovered that tawny owls have their own accents and dialects. They are not the only birds known to have regional accents. Chaffinches also have regional accents. In the 1950s a botany student called Peter Marler realised that chaffinches' songs differed from valley to valley in Scotland and the Lake District. He subsequently compiled dialect maps for chaffinches (and other songbirds) and concluded that the chaffinch learnt its song from its father. A similar study in the 1980s concluded that the reason that there were so many variations in a small geographical area were probably because the father's song which the chaffinch learnt would have had imperfections and birds do not always stay in the area where they were born, they move around. Studies of redwings in Norway, carried out by

---

18 In Proverbs and Popular Sayings of the Seasons, published in 1846.

The Wonders of the Wild Places

Tron and Tore Bjerke, revealed that the redwings' songs changed every sixth of a kilometre. A redwing singing near a border occasionally slipped into the neighbouring dialect. However, there was no hybridisation. Tron and Tore Bjerke concluded that this enabled redwings to return to the same territories every year with astonishing accuracy. Rupert Marshall, at Aberystwyth University studies corn buntings dialects and notes that the difference is over such a small distance that it is like people on opposite sides of the street talking in different dialects. However, corn buntings have declined massively in the last few decades, and this has had an impact on their dialects. The different dialects have lost their definition, resulting in a, "mish mash of songs." This poses a conundrum for conservation charities who are trying to save these birds. Rupert Marshall makes the point eloquently when he says, "If we succeed in saving the corn bunting, but they no longer exhibit the behaviour for which they were once so well known (i.e. singing in dialects), can we truly have saved them? Imagine a lion that no longer roared."

Greater- spotted woodpeckers are another bird which are often heard rather than seen in broadleaf woodlands. These are medium-sized black and white birds. Male and young woodpeckers have red on the neck or head, which sometimes makes them easier to spot as they undulate through the wood. Greater -spotted woodpeckers probe tree trunks, using their long sticky tongues to extract insects and larvae. You would think that all that drumming away on tree trunks would give the woodpecker a nasty headache, but fortunately they have shock absorbers in their beaks and skull. Greater- spotted woodpeckers are territorial so a good way to try and see one if you can hear one nearby is to strike a pebble rhythmically on a tree trunk. I have tried this method several times and have seen woodpeckers on occasion. Greater- spotted woodpeckers have bucked the trend of declining bird populations. Their population has actually increased by 300% in fifty years, possibly because of Dutch elm disease which led to a lot of dead wood, an ideal habitat for greater-spotted woodpeckers, and possibly because they can now be seen in gardens as well. Two years ago, I was again at the right place at the right time, this time sitting having my picnic in an ancient Yorkshire woodland in late Spring, when I saw a lesser -spotted woodpecker for the first time in my life. These birds are much smaller than their more common cousins, about the size of a sparrow, but are also black and white. Again, the males have bright red caps. Like their more common, larger cousins they also like ancient woodland. One of the best places to see them in the U.K. is the New Forest in Hampshire.

Other birds that you can quite commonly see climbing up or down tree trunks are nuthatches and tree creepers. Nuthatches are one of my favourite birds. They have blue-grey backs, orange breasts and a black stripe over their eyes. Like greater-spotted woodpeckers nuthatches particularly like mature oak woodland, often nesting in old woodpecker holes in trees.[5] Nuthatches normally can be seen descending head first down a tree, whereas the much plainer tree creeper is often seen ascending the tree.

The two birds of prey particularly associated with woodland are sparrowhawks and goshawks. It is usual in birds of prey for the female to be larger than the male, but sparrowhawks have the greatest difference of any British raptor, with the female being almost twice as heavy as the male. As the name suggests these birds mainly eat small to medium sized birds, although I have seen one take a pigeon in mid-air. They can synchronise their breeding period, so it coincides with when young songbirds are fledging. They fly quickly through woodland, tucking in their wings to get through tight spaces. Interestingly a female sparrowhawk is called a musket, which is how the firearm got its name.

In the 1950s and 60s organochlorine pesticides, most notably D.D.T., were used. These had a catastrophic effect on bird of prey populations because the birds and small mammals they had eaten had ingested the pesticide. Birds of prey had thinner, less viable eggs. Fortunately, these pesticides were banned in the 1980s and since then birds of prey numbers have increased, with sparrowhawks now regularly seen in gardens as well as woodland.

Goshawk numbers are still quite low though, although I have been lucky to see two in local woodland. These are similar in appearance to sparrowhawks but much larger. These birds tend to eat bigger prey, such as rabbits, corvids, and squirrels, and like to live on the edge of woodland although in several European cities, such as Berlin, they are now thriving in city parks too. In recent years scientists have recorded that male goshawks are getting smaller, making them more agile, whereas female goshawks are getting larger, so much so that they can now catch hares.

Threats to woodland

- **Undermanagement**

- **HS2**- According to the Woodland Trust, 108 ancient woodlands face destruction or damage because of the ongoing HS2 construction. The project is still in its first phase and already several ancient woodlands have been felled. Although HS2 claims that it is planting millions of trees to replace those lost they fail to understand that the trees they are planting will take hundreds of years before they provide the same benefits to nature as the ones they have cut down.
- **Tree diseases and pests**- including ash dieback, acute oak decline, dothistroma needle blight, oak processionary moth, and horse chestnut leaf miner. Many of these pests and diseases threaten our most valuable habitats, as I have described in this chapter, and the effect of them is devasting. For example, we risk losing 90% of our ash trees because of ash die back. To put that another way, that is 150 million mature trees and 2 billion saplings in the next 10-20 years, and that is just due to ash dieback. There are currently about a thousand pests and diseases affecting trees in the U.K. This will have a huge impact on our landscape and our biodiversity. For example, ash trees support 44 species which cannot be supported by other species of tree. If we lose these trees, we lose these species too. These diseases are prevalent because unfortunately at the moment there is a perfect storm of globalisation, imported trees, airborne threats, neglect of woodland, pollution, climate change, and an increase in animals, such as deer and grey squirrels, which damage trees. The latter makes it harder for the tree to use its natural defences, which I have described in this chapter, to kill the pests or fungi which are destroying it.
- **Air pollution** – see the Urban Jungle chapter for further details.

What is being done

- **Undermanagement** – Conservation charities, such as the National Trust and Woodland Trust, have started going back to more traditional methods of managing woodland in recent years. Coppicing has started to become a method of management again. This improves biodiversity and ensures a healthier woodland, hopefully allowing trees to live longer.
- **Tree diseases and pests-** The government is trying a three-pronged response of resistance, response and recovery, and adaption. This involves encouraging home grown propagation and the planting of native species, more research and innovation, increasing diversity of species and genetic variation (by selecting certain genotypes it might be possible to grow trees which are resistant to the pest), managing woodland better (a healthy woodland is more resistant to threats) and encouraging more natural regeneration of woodland (because naturally regenerated trees tend to be more resistant to the threats in their area). I am an Observatree volunteer. This is a joint project between a number of organisations including the Woodland Trust and Forestry Commission. It is a huge citizen science project. All over the Britain there are volunteers surveying local woodland to look for signs of pests and diseases and then relating the information back to scientists in order that they can get an idea of the scale of the problem and discover ways of combating it.
- **Pollution-** The government has vowed to ban petrol and diesel vehicles in the next 10 years, but it is not clear whether there will be the infrastructure needed for electric cars by then.

Why should we protect our woodland?

- They provide a habitat for numerous species, including some of our rarest species. Deadwood is also important, 20% of all known species in the U.K. rely on it.
- They absorb carbon. One tree can store 22 tonnes of carbon dioxide.
- They release oxygen.

- They reduce flooding and soil erosion by absorbing thousands of litres of stormwater.
- Being amongst trees is good for human beings physical and mental health. Trees release airborne chemicals called phytoncides to protect themselves against germs and pests. When humans breathe in these phytoncides it improves not only their immune systems but also increases white blood cells which kill tumours and viruses, as well as lowering blood pressure and anxiety and depression, and improving sleep.

What can you do?

- Plant a tree- Make sure that it is a native home-grown tree though. Also think about where you are planting the tree and whether it is likely to thrive in that habitat. Have a look on the Woodland Trust website for further information. The National Trust also have a tree planting scheme. At £5 a tree I have bought several as memorials to several relatives. Countryfile also have a Plant Britain campaign. Details are in the Appendix.
- Join a local group or charity and help maintain woodland. I am a Woodland Trust volunteer. More information can be found on their website.
- Take part in two large citizen science projects both organised by the Woodland Trust- Ancient Tree Inventory (to record ancient and veteran trees) and Observatree (looking for signs of pest and disease). Again, look at the Woodland Trust website for further information.
- Campaign about HS2- write to your MP.

Hedges and Edges

Following on from the last chapter, of course, trees do not just live in woodland. Hedges are also an important habitat for wildlife. These are the most widespread semi-natural habitats in the United Kingdom. According to D.E.F.R.A. [19] there are 402,000 km (249,791 miles) of managed hedges and 145,000 km (90,099 miles) of relic hedges (which are hedges which have not been managed and have therefore, in some cases, just reverted to being lines of trees.) This might seem like a lot, but actually between the end of World War Two and the 1990s the U.K. lost 121,000 km (73,186 miles.) of hedgerows. What is even more staggering is that in the 1970s the government was actually paying farmers to remove hedges in a bid for bigger fields for the over industrialised agricultural system we have today. This not only effected our agriculture and food production, and the look of our rural landscapes, but also had an effect on our wildlife. Like ancient woodlands, ancient hedges can be biodiverse and an important habitat for some of our rare species, including some of our rarest bats, hazel dormice, and the great-crested newt. An ancient hedge can support more than 600 plants, 1500 insects, 65 birds and 20 mammals.

So, what constitutes an ancient hedge? Like we saw with woodlands in the last chapter, this is not an easy question to answer.

The oldest hedges in Britain date back to the Neolithic period (4000-6000 years ago) and can mainly be seen in the South-West of England, especially in Cornwall and Devon. In Cornwall 75% of the hedges are ancient, whereas their neighbours, Devon, claim that more than a quarter of their hedges are over 800 years old. The reason why so many old hedges have remained in this part of the country is that this region was not affected by the centuries of Enclosure Acts which saw a huge shift in who owned the land in this country. As rural inhabitants lost the right to common land, and landlords charged higher rents for access to land which had once been common ground, many impoverished working-class rural

---

19 Department of the Environment, Food and Rural Affairs

people moved to urban areas. This can still be reflected today, with almost 84% of the British population living in urban areas.

The Romans planted thorny hedges around their fields and orchards, and after they left Britain in 400AD, the Anglo-Saxons continued to plant hedges. The Saxons had several words for hedge, including *hega* or *hege* which referred to a living boundary, as opposed to *gearde* for fence (which is where we get the word garden from). *Heag* was the word for a hurdle and tended to be used for a territorial boundary, whereas *haga* referred to an enclosure. They also gave us a compound word, *hegeraew,* which over the years became *hedgerow.*

At this time a lot of the land which these hedges surrounded were common land, farmed by all the people in the village, to provide food for the villagers. This open field system was replaced by a manorial system by the Normans, but it still involved communal land although now the people working it where doing it on behalf of the lord of the manor.

Systematic planting of hedges started in the 13th century because of the first enclosure[20] movement. This responded to an increasing population and more land required for agriculture and was halted by the Black Death which saw a huge drop in the population.[21] However, during the Tudor period, keeping sheep became more common because there was a drop in the population and sheep were far less labour intensive than arable land. Therefore, there were more Enclosure Acts as hedgerows were planted to stop livestock escaping. When out and about look for straight hedgerows near roads and paths. A lot of these will date back to the 16th and 17th century enclosure movement, which came to a head at the end of the 18th century. These had a huge effect, effectively denying the majority of the population access to the land. By the First World War there had been 5200 Enclosure Acts which had transformed 6.8 million acres in England and 1.7million in Wales. In the late Victorian period to the beginning of the Second World War there were further changes. Cheap imports of grain from North America, and increased mechanisation led to a reduction in arable land. Many miles

---

20 Enclosure refers to the physical boundary whereas inclosure is the legal term to convert common land into private land.
21 It is estimated that 30-45% of the U.K. population died between 1348-50 because of the Black Death.

of hedgerows were removed. This was accelerated after the Second World War when farmers were paid to remove hedgerows as there was a huge drive to increase yields from agricultural land.

Although you can get some idea of the age of a hedge by looking at a landscape – a hedge with a rounded bit near the entrance is probably medieval because it was designed to allow room to get the plough into the field- it is difficult to date an individual hedge. When I was a child, I used the Hooper formula, measuring thirty metres of hedge, and then counting the species – a century for each species spotted. However, it has been shown that this is not particularly accurate because there are regional variations. Where I live in the North of England, it was traditional to use fewer species to form a hedge, whereas a lot of enclosure hedges were traditionally just one species – hawthorn. Similarly, as we strive for greater biodiversity, many hedges planted in the last thirty years have contained a variety of species. If a hedge contains many fruit trees it is likely to be an older pre-enclosure hedge, because that used to be a good way of getting food throughout the year, but even this cannot be totally relied on. In short, it is really difficult to date a hedge.

One way is to look at the species within the hedgerow and the hedgeback (the bottom part of the hedge). Species such as wild service, spindle, dogwood, and hazel often indicate an older hedge. Spindle is seen more commonly in the South of England. It has bright pink flowers in the spring, which are an excellent source of nectar and pollen for pollinators such as bees. As the name suggests it was used to make spindles for spinning, knitting needles and pegs; as well as providing high quality charcoal for artists. Be careful because the leaves and fruit are toxic to humans, but valuable habitats for moths including the spindle ermine and the holly blue butterfly.

The most common tree in hedgerows in the United Kingdom is hawthorn. This is reflected in the numerous British place names with thorn in their names. It gets its name from its red fruit, or haws, which is another word derived from the Saxons' many words for a hedge. There are two species in the U.K. – Common and English hawthorn. [22] Hawthorn is a hardy tree, resistant to disease and long living. It is also a host for many

---

[22] English hawthorn just grows in an area below the M62 down to Bristol and Brighton.

different species. Hawthorn wood was used in the teeth of mill wheels, dagger handles, hammers and walking sticks, and its thorns were used as needles and fish-hooks.

It is also known as May blossom because its flowers are prevalent in that month and were once used as confetti because it was traditional for people to get married in Spring. Although they are a pretty flower do not be tempted to pick some and put it in a vase or use it on your wedding day because the blossom contains a chemical called trimethylamine, one of the first chemicals produced by decaying animal tissue, so it was thought to be unlucky to bring it into the home.

Hawthorn has always been associated with sacred sites. It was thought to have been the basis of Christ's crown of thorns and the Glastonbury Thorn in Somerset unusually flowers in December. It is said to derive from Jesus's Great Uncle, Joseph of Arimathea, who is said to have visited Britain and planted his staff. Hawthorn has always had a strong association with fairies and Gaelic folklore claimed that hawthorn marked the entrance to the *other world*. Indeed the 13th century poet, called Thomas the Rhymer, claimed that the Queen of the Fairies had met him at a hawthorn and then led him into the Underworld, where they had got married before he re-emerged seven years later.

Another species commonly found in hedges is blackthorn. Like hawthorn it makes a good hedge because it is impenetrable and provides purple-blue fruit, called sloes, in the Autumn. A good way of telling them apart is that blackthorn's white blossom is one of the first things to bloom in March and April, before the leaves have come out, whereas hawthorn's leaves come our first followed by its blossom in May. It is home to the rare black hairstreak butterfly and provides wood for witches' wands and staffs.

The wayfaring tree can also be seen in hedges. It is so called because it tended to be planted close to paths and old roads. Its bright red berries are a valuable food source for many birds and insects. In Europe, its wood was used to make arrows.

Another tree which commonly appears in hedges, especially older Enclosure hedges, is crab apple. Interesting the many species of domesticated apple trees which we have in Britain are not related to the European crab apple; actually, thought to descend from apple trees in the

The Wonders of the Wild Places

Tien Shan region of North-West China. They are thought to have been brought over to the U.K. in the Middle Ages from Continental Europe. However, crab apples are used in commercial orchids because they make excellent pollination partners for cultivated apple trees. Like cultivated apples crab apples are edible, although much tarter, and can be used to make jelly or as a natural pectin for making jam. It is also a host for many species including mistletoe, lichens, and many species of moths; badgers, foxes, voles, blackbirds, and crows love eating crab apples, and the blossom is an early source of nectar for many pollinators.

One of the rarest species, so rare that it is protected by law and its seeds are in the Millennium Seed Bank at Kew, is the Plymouth pear, so called because it was first discovered in Plymouth in the 19th century (and even today can only be seen in Plymouth or Truro.) It has smaller, rounder pears than cultivated pear trees and smells very unpleasant. It has a very unusual in-built control mechanism, self-incompatibility, which prevents it breeding which explains why it is extremely rare.

Another thorny species often found in hedges is dog rose whose rose hips provide such a good source of Vitamin C that during the Second World War, when imports of citrus fruits were not possible, the government advised people to eat rose hips to prevent Vitamin C deficiency. Rose hips are said to be a natural diuretic and laxative, as well as being used by herbalists to treat kidney infections, arthritis, urinary infections, and colds and fevers.[23]

An interesting thing which you can sometimes see on dog or field roses, which I saw for the first-time last year, is a small fuzzy red ball known as a robin's pin cushion. However, despite the quaint sounding name this is actually a gall created by a gall wasp which lays it larva on the rose. Each larva resides in its own chamber within the gall, emerging as adult wasps in the Spring. Gall wasps are unusual because they are parthenogenetic, meaning they do not need male wasps for reproduction so fewer than 1% of the wasps are male.

Bramble is in the same family as dog rose – the Rosaceae family. In Autumn its blackberries provide food for many birds and animals as well

---

23 Always seek medical advice before trying natural remedies.

as humans (apple and bramble crumble comes highly recommended) and in the Spring its pretty white flowers provide an important source of nectar, especially for brimstone and speckled butterflies.

Holly is another species which makes a prickly, impenetrable hedge which is good at stopping livestock, but also an excellent habitat for various species. It was often planted in deer parks and medieval hunting estates which is why it regularly appears in many placenames. Holly trees can live up to 300 years and older trees have smoother leaves, although with most holly trees the prickly leaves tend to be at the bottom of the tree to deter pests. Scientists have discovered that in environments where there are lots of grazers, such as deer, holly trees grow pricklier leaves under the height of 98 inches (2.5m), which is the average reach of a red deer. They are hermaphrodites, meaning that the male and female flowers are on the same tree. In Autumn and Winter, they have bright red berries, which are enjoyed by many birds including redwings and fieldfare which fly here from Scandinavia and Siberia in the Autumn. The thorns and berries of a holly are said to symbolise Jesus' blood on his crown of thorns, and prior to Victorian times holly trees were used as Christmas trees. It is associated with Christmas, and appears in a lot of traditional carols, because the Romans associated it with Saturn, their God of Agriculture and Harvest, and celebrated the Festival of Saturnalia, which was celebrated at the Winter Solstice, just before Christmas. During this festival Romans bedecked their homes with holly. Unlike hawthorn, it was thought of lucky to bring holly into the house because it was a fertility symbol and it was believed that it warded off the Devil, witches, goblins, and evil fairies. It was also planted near houses (I regularly see it near old farmsteads on the North York Moors) because it was thought to prevent a house being struck by lightning. Interestingly, in recent years, scientists have discovered that there is actually some truth in this superstition because the prickly leaves act as lightning conductors. Holly often appears in urban hedges and gardens because Victorians liked it because it is tolerant to pollution. Holly has always been a popular wood for woodturners and carpenters because it is a nice colour and easy to engrave. Traditionally it was used to make whips for ploughmen and horse-drawn carriage drivers because it was associated with control.

The Wonders of the Wild Places

Many plants like hedgerows, either living along the bottom of the hedgerow or on it. Flowers such as bluebells, the pretty white flowers of greater stitchwort, primroses (literally the first rose of spring), Jack-by-the hedge, cow parsley, hemlock, common valerian, comfrey, betony and various species of speedwell are just some of the 33% of British plant species which live on or near hedgerows. Primroses were used by the Romans to cure malaria and jaundice. If drunk as a tea it can be used to treat arthritis and migraines, an infusion of the flowers can be used to cure insomnia, and if made into an ointment it can be used to treat chilblains. Jack-by-the-hedge is also known as garlic mustard. Its leaves smell of garlic and can be eaten in salads and can be used to flavour fish and meat. Cow parsley is a pretty plant which can be seen alongside hedgerows and rural lanes in late Spring and is an early source of pollen for many pollinators. For centuries different counties, and even often different towns and villages, had different names for plants and birds. One of cow parsley's other names is Queen Anne's lace because the delicate lacy flowers looked like the lace worn by Queen Anne and her ladies- in-waiting. Cow parsley is very useful for people like me who represent a tasty treat for insects because it repels midges. Cow parsley is sometimes confused with the much more dangerous hemlock. Hemlock is taller than cow parsley and has a much bigger cluster of umbel white flowers, which are very pretty, but do not touch the red stem because it gives a really nasty burn. Hemlock is also extremely poisonous and even ingesting a tiny amount is fatal. In contrast common valerian, which has a white flower, and red valerian are very useful plants. For centuries valerian has been used to treat hysteria (and it is where we get Valium from) and is also a useful sedative and painkiller. Comfrey is another pretty and useful plant. Its name is from a Latin word *symphytum* which means to *grow together*. It contains a chemical which increases the rate which the body mends bones, ligaments and tendons and it contains an alkaloid allantoin which repairs and replaces damaged cells of damaged connective tissue, so can be used to treat ulcers. It was introduced to the United Kingdom by knights returning from the Crusades, who had used the plant to treat wounds incurred on the battlefield. It is also used to treat stings and bites and skin problems. However, like a lot of British plants, it is toxic if ingested. It can also be used to create fertilizer, just put the leaves in rainwater and leave for a few weeks. It is effective but beware, the smell is akin to rotten eggs.

Stitchwort not only cured stitches (which in Medieval times people thought were caused by elves) but also mends broken bones. Betony is a small plant with red-purple flowers which was considered a holy herb in the Middle Ages and was thought to repel witches and the Devil. It was introduced to this country by the Normans and is a very useful herb- Emperor Augustus' physician claimed it cured forty-seven disorders- and can be used to staunch bleeding, heal tissue, improve concentration and memory, as well as a sedative and antiseptic. It can be confused with a smaller but similar plant, self-heal. As the name suggests self-heal was deemed to have such excellent healing properties that someone could heal themselves without the need of a trained physician. It is antiviral and an antioxidant and is used today to treat fevers and colds, herpes, AIDS, and thyroid problems. Scientists are also carrying out studies to see if it can be used to treat high blood pressure and diabetes. It was also known as the carpenter's herbs because it stems blood flow and helps wounds heal.

Hedge woundwort which has similar colour flowers but is a much taller plant. Any flower with the suffix -wort means it is thought to have herbal properties and the leaves of woundwort can be used to clean a wound and stem bleeding. There are various species of a pretty, delicate blue flower called speedwell, (so called because it literally speeds you well on your journey, bringing you good luck and health) which can be seen along grass verges, hedgerows, gardens, and wasteland. Many species of speedwell were introduced to the U.K. from Asia by the Victorians, but our most common native species is germander speedwell, whose name derives from the Greek meaning *oak on the ground*. In Ireland it was sewn into clothes to as a charm against accidents.

Cuckoo pint is another plant which can be found near hedgerows. It is also known as Lords and Ladies because the Elizabethans used them to starch their ruffs, although it caused their hands to blister if they touched it. They are strange looking flowers. The Victorians thought they were indecent because their flowers reminded them of the male and female reproductive organs. 'Pint' is derived from the Old English word *pintel*, meaning penis. The purple 'poker', which so upset the Victorians, has a nifty trick. It warms up and emits a smell of urine, which lures flies because they think it is rotten food. Instead, the flies get imprisoned by a ring of hairs pointing downwards. When the female flowers are pollinated and the male flowers shed pollen, the hairs whither and the flies are freed. Covered in pollen the flies then fly to another cuckoo pint where the process is repeated. I would have thought that after a while the flies would cotton on, but it does not seem to happen.

Meadowsweet is another plant that can be seen growing on grass verges. It is one of the plants that contains salicylic acid so has many medical uses from curing indigestion, to treating coughs and fevers, to relieving arthritic pain and headaches, to treating cystitis and kidney problems. As if that was not enough it is also an antiseptic and diuretic. It was said to give someone second sight and is also known as bridewort because it was part of Tudor bridal bouquets. It was used to flavour beer, before hops were introduced to Britain, and the dried leaves were smoked like tobacco.

Saint John's Wort grows in hedgebacks and grass verges, as well as woodlands. One of its other names is *fuga daemonium (Devil's Flight)* because it was believed that the Devil hated this plant so much that he tried to destroy it with a dagger or needle, causing red spots to appear on the plant on the 29th August- coincidentally the day on which St John the Baptist (after whom the plant is named) was beheaded. The red spots are also said to be the Saint's blood, which is said to strengthen the plant, protecting it against the Devil. The story behind the plant might differ, but it was generally agreed during the Middle Ages that this was the most powerful plant in the fight against evil. It is also a very useful medicinal plant. It is probably best known as a treatment for depression, but it is also a sedative and a painkiller (it is used to treat conditions such as rheumatism and headaches), as well as being made into a salve and used to treat burns, sores, and nerve pain. It can be mixed with oil to treat chest complaints. Knights during the Crusades used it to treat open wounds. When mixed with alum it makes a yellowish-red dye which is used in the production of Scottish tartan.

Flowers can grow in the hedgerow itself. Hedge bindweed has a white flower, but it can be a problem because it smothers other plants.

Also look for field bindweed which grows on the ground and has pretty candy-pink and white flowers. The sweet-smelling honeysuckle is one of my favourite flowers. It is often seen growing around the doors of old country cottages because it was thought that it brought good luck to the home and stopped evil spirits from entering the premises. It is also a very valuable habitat and food source for many birds, mammals, and insects, including some of our rarest ones. Dormice like living in it, and the nectar-rich flowers support the rare white admiral butterfly. The flowers are rich in salicylic acid (perhaps better known as aspirin) so are used to treat headaches, colds and flu, fevers, and rheumatism, as well as bronchial problems, such as bronchitis. The stems were used in the textile industry as well as for binding.

Ivy can often be seen in hedges and is very prevalent in the hedges near where I live. Mature ivy has a pretty flower and provides a valuable habitat for fifty species, including the rare golden hoverfly. It used to be thought of as parasitic because it can often be seen growing on trees, but actually it has its own root system and does not damage the tree. It has specialised hairs which help it stick to trees and buildings and allow it to climb. Like holly, it was seen as a sign of fertility, and it was traditional for a priest to place ivy wreaths around the necks of newly married couples and, even today, many bridal bouquets contain a sprig of ivy. It also symbolised intelligence which is why poets, writers, and philosophers are often depicted with wreathes of ivy around their necks.

In the North of England, where I live, stone walls are more common than hedges, although there has been a reduction of these in recent years too[24]. Some of these stone walls probably date back to the Neolithic period, about 4000 years ago. These are also valuable habitats, providing a home for songbirds, small mammals, and reptiles, as well as lichens and mosses. They have their own microclimates, and the nooks and crannies provide important habitats for wildflowers. It has also been shown that bats and bees use linear structures, such as stone walls, to help them navigate.

Walls and hedgerows are important habitats for songbirds and support 80% of the U.K.'s woodland bird species. As a child I used to love looking

---

24 Although farmers now get grants for building or repairing stone walls

at birds' nests in hedges. It is possible to tell which songbird has made it by looking at what the nest is made of[25]. With one of our smallest and best-known birds, the wren, it is the male that makes the nest. He actually makes four to six nests, and then parades the female wren round them. If she decides she likes one, he breeds with her. However, he is unfortunately promiscuous and can breed with several females in one year. It probably is not a co-incidence that the name wren is from the Saxon word *wroenna*, meaning *lascivious*.[26]

The eggs can also give you a clue to the species of the bird. Birds such as blackbirds lay blue eggs, possibly to protect the unhatched chicks from damaging ultraviolet light. Most birds try to camouflage their eggs to protect them from predators so eggs lain in hedgerows are normally green or blue, whereas birds that lay their eggs on bare ground normally lay brown eggs; whereas birds, such as woodpeckers, which lay their eggs in holes in trees often have white eggs so that they can see their eggs in the dark. Scientists have discovered that the flying ability of the bird also affects the shape of the egg, with birds that fly faster laying more asymmetrical- shaped eggs.

Grass verges are alongside many hedges, as well as alongside many of our roads. It might surprise you, but 45% of the wildflowers found in Britain can be found on grass verges, including 29 out of the 52 wild orchids found in the U.K. including very rare orchids, such as the lizard orchid. Red clover, Lady's bedstraw, silverweed, and giant hogweed are some of the species commonly seen on grass verges. Red clover has many health properties. Middle - aged ladies use it to alleviate the symptoms of the menopause, and it is also thought to have blood thinning properties which scientists are studying to see if they can prevent blood clots. Ever since Celtic times finding a rare four- leaf clover is thought of as being lucky. It was believed that if a four-leaf clover was placed under someone's pillow they would dream of their true love. It was thought to be a magical plant and was said to grow only where elves lived. It was worn to ward off witches and it was believed that if it was worn on

---

25 Do not remove the nest, the bird might need it for next year.
26 feeling an overt sexual interest or desire.

clothing the wearer would be able to see fairies. It gets its name from the Latin word *clava* meaning 'a club' because of the three-knotted club of Hercules, which is the symbol you see on playing cards. Lady's bedstraw is one of several bedstraw species in the U.K. and can be identified by its yellow flower. It gets its name because in the Middle Ages, when there was a cult surrounding the Virgin Mary, making her more important that Jesus, it was alleged that the Virgin gave birth to Jesus on a bed of Lady's bedstraw and bracken. The bracken refused to acknowledge Christ as the son of God and lost its flower, whereas Lady's bedstraw acknowledged Jesus as the son of God and was rewarded with a golden flower. Even before Christ it was an important flower in Norse mythology associated with Frigg, the Goddess of Married Women, who was said to help women during childbirth. Scandinavians used it as a sedative for ladies in labour, and in the U.K., it was used to stuff beds, especially the beds of expectant mothers, although this might also have been because it repels fleas. It was also used as a substitute for rennet in cheese making because it contains a protein which curdles milk, and the yellow flowers were used to dye the cheese. Silverweed has silvery leaves and buttercup-like yellow flowers and was used by Roman soldiers to pad their shoes because it prevents sore feet. As a keen fellwalker I have used it myself and can recommend it for footsore walkers. Giant hogweed[27] originates from the Caucasus mountains and Central Asia and was introduced to the U.K. by the Victorians, who had a passion for plants from all over the world. It grows very tall, up to sixteen feet (five metres), but should not be touched because its sap contains furocoumarin which makes skin extremely sensitive to sunlight and causes severe blistering which can reoccur over months and years if the area is exposed to sunlight.[28]

Burdock is another plant you might see on grass verges. Anyone who has tried to remove a burr from their clothes might not be surprised to read that they inspired George de Mestral to invent Velcro in the 1940s. Burdock is used to treat rheumatism and joint pain, staunch internal and external bleeding and can be applied to sores and ulcers. Chickweed is another common plant found on grass verges. As the name suggests it was

---

27 There is also hogweed which is less toxic but can still cause skin irritation.
28 If you do accidentally touch giant hogweed seek urgent medical advice.

used to feed farmland birds. It is used to treat obesity because it contains saponins which inhibit intestinal absorption of fats and carbohydrates.

Nettles are another plant which are often seen on grass verges, and hedgebacks. I doubt many people would list nettles as one of their favourite plants, certainly if you have ever been stung by one. They get their name from the Anglo-Saxon word *netel* which is related to the word *noedl*, meaning needle, probably because it is painful if you catch yourself on either a needle or a nettle. Incidentally, rhubarb leaves, mint and rosemary are allegedly effective in taking the pain away from nettle stings.

Most people will be surprised to read that there is far more to the humble nettle than meets the eye. They make a very strong natural fibre which was used to make clothing in the Bronze Age, although it has been calculated that it takes 40kg (just over six stone) of nettles to make one shirt. Nettles were used during World War Two to make ropes and parachutes. During the 18th century nettle fibre was used to make sheets and tablecloths in Scotland and paper in France.

Nettles contain serotonin, a neurotransmitter which affects our mood, and histamines. The Germans used them to treat arthritis and it has been shown that nettles contain a chemical which protects the synovia fluid in joints. It can be drunk as a tea and used as a diuretic and to treat respiratory problems. It has also been shown to lower blood sugar levels and is high in iron, so it is used to treat anaemia.

As if that is not enough nettles can also be used as a dye, and if soaked in rainwater for several days they can make a good natural insecticide and fertilizer. Nettle juice was used to curdle milk, and, before paraffin, nettles were used as a light-source in lamps.

Grass verges are not just a good place to see common plants, such as nettles. Incredibly, some of the rarest plants in the U.K are found on grass verges. These include crested- cow wheat which is just found in Cambridgeshire; sulphur clover which enjoys grass verges in the East of England, velvet Lady's mantle which is reduced to six plants in the North of England, and tower mustard which can be found in small numbers on grass verges in Gloucestershire, Suffolk, and Hampshire. The rarest one is the fen ragwort. Only one plant is known to exist alongside a busy A- road in the East of England. Of the 700 plus species found next to U.K. roads

12% face extinction. For years, these plants were under threat because of overzealous grass cutting by local authorities. This resulted in a loss of a fifth of the plant species near Britain's 315,500 miles of road verges. However, there is good news. The Highway Agency, in conjunction with organisations such as Plantlife, have published new guidelines suggesting that verges should only be cut twice a year. If this is done it is thought that it could result in a staggering 400 billion more flowers. To put it another way if all local authorities follow the guidelines an area the size of Nottinghamshire could see hundreds of species of wildflowers thriving. Several local authorities, partly due to budget cuts, have adopted this advice. In saving nature, they can also save themselves hundreds of thousands of pounds per year. Win, win. This is not only good news for our wildflower species, but also for our human population because it brightens up towns and cities and provides valuable wildlife habitats and corridors which is vital if we are going to radically improve our natural environment in the next decade. It is also something important to think about when designing new roads. The Weymouth relief road is only four and a half miles long, but since it was opened in 2011 more than 50% of British butterfly species have been recorded alongside it.

As well as cutting less Natural England has also suggested that local authorities and the Highway Agency consider the effect that pollution, grit and de-icing salt, pesticides and fertilizers have on these special habitats.

Animals and birds also live alongside some of our busiest roads. A bird which I regularly see as I drive along is one of the smallest birds of prey, the kestrel. These birds are also known as windhovers because of their habit of hovering above their prey. Their numbers are linked with those of their main prey- voles. Kestrels can see in ultraviolet light enabling them to see the urine trails left behind by small rodents. They also have twice as many photoreceptors as we humans do which allows a kestrel at the top of a 59ft (18m) tree to see a tiny invertebrate on the ground really clearly.

Hedges and edges are also a good place to see mammals. Stoats and weasels look similar, but stoats can be differentiated because they are bigger than weasels, which are the U.K.'s smallest carnivores.[29] Stoats

---

29 Female weasels only weigh 1.7 oz/50g

move with a bounding gait whereas a weasel moves quickly close to the ground. Stoats have a black tip to their tail.

Stoats are great climbers, swimmers, and runners. They can reach speeds of 20mph, an incredible speed for such a small animal, and they can cover five miles during one hunt. They also have excellent senses, especially their sense of smell. Amazingly they can tell the gender, health, and age of their prey just from its smell. They can also determine whether their intended prey is pregnant, thus making them slower and easier to catch, just by their smell. However, rodents, such as voles, have evolved to enable them to shut down their reproduction system if they smell a stoat, thus making it harder for the stoat to detect them.

Weasels prey on voles and mice, which make up 60-80% of their diet. As they are small creatures they require a lot of food for energy and must eat about 25-33% of their own body weight every day to survive. Like stoats they have big territories of 4-8 hectares.

Why are hedges and edges important?

They provide wildlife corridors. One of the problems faced by wildlife in the U.K. is that there are pockets of many different habitats (in the form of nature reserves, national parks, areas of outstanding natural beauty, and sites of special scientific interest, just to name a few) but it is difficult for wildlife to move from one to another. This creates isolated communities and also means that if they are struggling in one area it is harder for them to move to another area where they might be able to adapt. It also can make it harder for wildlife to travel for food, water, and shelter. Stone walls and hedgerows and other linear structures, such as railway lines and rivers, help provide protection for wildlife and allow them to travel safely. They link habitats and strengthen gene pools.

They provide valuable habitats not just for hundreds of insects, birds, reptiles, mammals, and plants, but also for over 125 priority species[30] including turtle doves, ciri buntings, brown hairstreak butterflies and greater horseshoe bats.

---

30 Our most at risk species

They trap pollution in urban areas (see the Urban Jungle chapter)
They prevent soil loss from adjacent fields. Hedgerows and stone walls can act as important barriers, preventing water run-off, reducing wind erosion, and regulating water supply. A one-hectare field with a 50m hedge at the bottom can store 150-375 cubic metres of water during rainy periods, particularly if it's a clay soil or the land is farmed organically. The deep roots of the hedges can also prevent flooding by removing excessive water.
They aid navigation. Birds, bats, and other species use hedgerows and stone walls as landmarks to help them navigate home and to find food. To my amazement I witnessed this last year when I saw a cuckoo flying over the River Ure in North Yorkshire.

Threats

Only 41% of hedgerows are in favourable condition for wildlife. The best hedgerows need to be slightly overgrown although if they are not managed at all they become relic hedges, mainly just rows of trees, which are also important habitats, but wildlife corridors work less well if there are lots of gaps. Cutting the hedge too often is also bad for nature because it prevents things from growing and destroys habitats. Although it is illegal to cut hedges whilst birds are nesting, in the Spring and Summer, I regularly see farmers near me cutting hedges in the middle of April. Machine trimmed hedges are of little biological interest.

Like woodland, hedgerows also suffer from disease and pests but some species, such as hawthorn, are more resistant.

There is some legal protection in the form of the Hedgerows Act 1997, but this does not include stone walls and just covers England and Wales. The Act is of limited use because it just protects hedges which are at least 20m(66ft) long and have some importance, such as archaeological interest, marking the boundary of a pre-1600 estate or part of a field system which predates the Enclosure Acts. Hedgerows which support more than seven species[31] are also regarded as being important. There is a list of fifty species on the Red List (so at risk of extinction). However,

---

31 There are some regional variations, there are less species in the North of England

in reality most hedgerows are not covered by this Act and there has only been one prosecution for removing a so-called important hedge.

Hedgerows which are adjacent to arable fields tend to have less species because of the effects of pollution from fertilizers.

What is being done?

Organisations such as the National Trust, Wildlife Trusts, and RSPB are creating and preserving wildlife corridors. This is already having a positive effect. The National Trust are major landowners in the U.K. They own a lot of farms which they want to farm in a more nature friendly way, including restoring and planting hedgerows and stone walls. At the Sherborne Park Estate, they have created and preserved hedgerows which now link up untidy field margins and woodland. This has led to eleven species of bats living in the area because they are able to use the hedgerows to provide them with food and for navigation.

Britain has left the European Union and the various schemes which it provided for encouraging farmers and other landowners to preserve and create habitats on their land. The government plans to provide Environmental Land Management schemes which will reward farmers if they prevent flooding, plant woods, and help wildlife. This includes protecting stone walls and expanding and planting hedges, as well as reducing the use of pesticides. However, at the moment, these plans lack clarity and conservation organisations are critical because the plans are not due to come into force until 2028, which they argue is too late.

Plantlife are encouraging Councils to mow verges less often. The charity's Dr Trevor Dines claims that if verges are managed better there would be more wildflowers covering, "...an area the size of London, Birmingham, Manchester, Cardiff, and Edinburgh combined ..."

What you can do

- Do not cut your hedges too often- once in about February and again in about September is enough.
- If you have a straggly hedge, try layering it. This is when trees are partially cut then bent over. It should be done every fifteen years. There are different techniques depending on which region you are in. Once it is layered just trim the hedge every two or three years to allow time for fruits and blossom to flourish. The National Hedgelaying Society can provide more information, as can charities such as the National Trust. There are also many videos on YouTube.
- If you have not got a hedge, plant one. Think about the best species to use. Plant native trees which are more resistant to pests and pathogens. Look at the Woodland Trust's MOREhedges scheme for advice and funding.
- Volunteer for a conservation organisation and help create wildlife corridors.
- Do not use pesticides and fertilizers.

On the Farm

According to a recent study at Sheffield University about 70% of the land used in the U.K. is used for agriculture, with about 56% of this being arable[32] land and pasture. This includes a very important habitat – meadowland. John Lewis-Stempel, in his wonderful book *Meadowland* describes it as, "...a place where grass and flowers are grown for hay." Traditionally land was managed on an annual cycle with the grass and flowers cut at the end of Summer and the area being grazed in the Winter. Prior to the Second World War every parish in the country had at least one meadow, but then the war put pressure on farmers to grow more to feed the British people and the government instructed landowners to plough meadows and turn them into arable fields or land for livestock. Consequently 97% of our meadows have been lost, with only 26,000 acres of lowland meadows and just over 2% of upland meadows remaining.

Alongside this another important habitat – grasslands- many of whom date back to the Bronze Age, has also been lost by the same amount. This has not only had an impact on the way our landscape looks but also on our wildlife. According to Dr. Trevor Dines at Plantlife few habitats in the U.K. can match meadowland and grassland for biodiversity. More priority species, which are species which are under threat, are associated with grassland than any other habitat, and meadows are incredibly biodiverse with forty species recorded per square metre. Meadows can also mitigate the effects of flooding, and meadowland and grassland can store carbon and improve soil quality.

Many mammals depend on these habitats including hares and rabbits. Hares and rabbits can be differentiated by their size (hares are bigger), the way they move (hares run rather than hop) and hares have black tips at the end of their long ears. Although they are thought of as being endemic to the U.K. it is thought that hares were introduced to the Britain by the Romans, or perhaps even earlier, whereas the Normans introduced rabbits

---

32 Used to grow crops

for food. Rabbits live in burrows whereas hares just make forms, usually in grassland. Hares can run at about 45 mph which enables them to escape most predators although foxes eat about 70% of young hares annually. Rabbits and hares have incredible almost 360-degree vision and rabbits can rotate their ears 180 degrees (and hares 270 degrees) so that they can pinpoint the exact location of the sound. Hares and rabbits, like a lot of British mammals, are crepuscular, which means they are best seen at twilight and during the night, although they are also often spotted during the day.

Rabbits are 'ecosystem engineers.' They are selective grazers which helps wildflowers, and their burrow- building and earth-scratching also help plants to germinate. The burrows also create 'mini mosaic habitats' which are wonderful places for rare flowers and invertebrates to thrive.

Hares and rabbits both feature a lot in culture, legend, and literature. Both species are associated with fertility and resurrection and in the Middle Ages hares were often given as love tokens. In Cornwall the sight of a white hare signified an incoming storm, whereas in Cambridgeshire if a hare was seen running down the street it was a sign that a fire would break out. Perhaps because they tend to be seen on farmland, hares were also associated with the harvest. The last sheaf of wheat to be harvested was known as the hare and the last cut was known as killing the hare.

The symbol of three hares chasing one another in a circle can be seen in churches, mainly in the South-West of England and is thought to possibly be connected to the Holy Trinity, although it appears in many other religions. The earliest example of it has been found in a Buddhist cave in China and has been dated back to between 581-618 AD. The symbol can be traced along the old Silk Road from China to Britain.

The tradition of carrying a rabbit's paw for luck dates to 600BC and is seen in Eastern and Western cultures. It was believed that carrying a hare's paw in the right hand prevented rheumatism and helped actors overcome stage fright.

The phrase *mad march hare* was first used in a 16th century poem[33] but was made popular following the Mad March Hare in Alice in Wonderland.

---

33 Replycacion by John Skelton 1528

The idea probably derives from hare's behaviour during the mating season in March when female hares stand on their back legs and literally fight off the advances of amorous males. It is an incredible thing to witness.

There are many famous hares and rabbits in popular culture, including Beatrix Potter's Peter Rabbit, Brier Rabbit, Little Nutbrown, and the rabbits in Richard Adam's Watership Down. However arguably one of the most famous rabbits is the Easter Bunny. It was thought that this rabbit had its origin with an Anglo-Saxon Goddess called Eostre, but this is controversial. The only evidence that Eostre existed is a mention of her by the famous Anglo-Saxon writer, the Venerable Bede. He said that Eosturmonath, or April[34] as we know it, was named after her, but historians dispute this because Anglo-Saxons normally named their months after agricultural and meteorological events rather than Gods. No images or carvings of Eostre have been found and she is not mentioned anywhere else. However, there was a Germanic Goddess called Ostara[35] who was the Goddess of Spring. In the 19th century, Jacob Grimm, of fairy-tale fame, surmised in a book about German mythology that Eostre must be a local version of Ostara, despite not having an evidence to back up the claim. In 1874 another German folklorist, Adolf Holtzmann, speculated that the Easter Hare, or Bunny as we tend to think of it in the U.K., was associated with Ostara. He claimed that she transformed a bird into a hare, which promptly laid coloured eggs for her. Holtzmann linked this with the tradition of giving Easter eggs. There is evidence that during the 19th century German children received toy hares and rabbits at Easter and were told the legend about Ostara and her egg-laying hare. Other people think that the Victorians were just desperate to legitimise traditions by trying to claim they had pagan roots. We will never know whether Eostre or Ostara existed, but it is an interesting story.

Sadly, hares and rabbits are struggling at the moment. Since World War Two haymaking has been replaced by silage production which only needs one species of grass, whereas hares need lots of different plants.

---

34 This is the Roman name for the 4th month of the year
35 Or Austra as she appears in some texts. Several inscriptions have been found referring to her in Germany.

The Wonders of the Wild Places

Another agricultural change which has affected these beautiful animals is the switch to planting crops in the Autumn, rather than the Spring when the hare needs more food because it is breeding and bringing up young. Farm machinery, such as combine harvesters, and the use of poison on farms, for reducing vermin, has also contributed to a decrease in hare populations, declining by 80%. It is feared that climate change will also affect them because they struggle to breed in wet conditions and our Springs are predicted to become wetter.

Recently it has been reported that rabbit and hare populations might be severely affected by a disease I last witnessed during my childhood- myxomatosis. This is a terrible disease which was shockingly deliberately introduced into Britain in the 1950s to reduce rabbit populations. As a child in the 1970s and 1980s I often came across rabbits with the closed eyes and swollen heads, two classic symptoms of myxomatosis. It destroyed the rabbit population, reducing it by a catastrophic 99%. Until 2018 it was thought that this disease only affected rabbits, but more than a thousand hares have been recorded with those symptoms since then and it is feared that the virus might have jumped species. Sadly this is not the only virus affecting rabbits and hares, several others have reached here from Europe. The worst one is rabbit haemorrhage disease virus type 2, which came to the U.K. in 2010 from France. It kills young rabbits. Scientists do not think that rabbit populations will recover. It is an even worse future for hares because scientists fear that hares will become extinct in the British Isles in the next decade or so, they are already locally extinct in some parts England. The thought of not seeing these beautiful, mystical creatures darting across a field in Springtime is heart-breaking.

In a field or meadow you might also see the sign of another British mammal, although you probably will not see it because it spends most of its life underground. Moles get their names from the Saxon word *moldwerp*[36] which means earth mover. This is a good name for them because these small mammals move a lot of earth- 10kg (1.5 stone) an hour. They dig an extensive network of tunnels. Sometimes generations of moles will use the same tunnels. At the end of the tunnels they dig chambers to sleep in

---

36 They are still known as this in some rural areas.

and give birth in. They also have a chamber to store worms. They keep the live earthworms immobile by biting off their heads. Up to 470 worms have been found in a chamber. As moles need to eat their own bodyweight in earthworms a day, it is important that they have plenty of worms.

Moles are not sociable animals so that there will only be about three to five moles in an average acre.

Shrews are another small mammal which can be seen in meadows and grasslands. Like weasels, shrews need to eat a lot each day, their own bodyweight in fact, to stay alive. Their saliva contains a poison which allows them to immobilise and kill their prey, which consists mainly of slugs and woodlice. It is rare that shrews are eaten by mammals because they have a gland on their flank which emits a foul smell. However, they are a favourite food source for raptors because birds do not have a sense of smell.

Meadows also support a wide range of plant species. Cowslips, from the Old English word for cowpats, is commonly found in meadows. They are antioxidants, anti-inflammatory and antispasmodic and they can also be used as sedative, and can be placed under a pillow to cure insomnia, amnesia, and asthma. As are meadow buttercups, which are a good indicator of old grassland. They are actually not very good for meadows because they contain a substance which removes potassium from the soil and inhibits Nitrogen- fixing bacteria, thereby causing other flowers to die, and allowing them to take over. Cows avoid them because they play havoc with their digestive system if eaten raw. Until the 18th century they were known as crowfoot. They might have got their current name because it was believed that if you rubbed buttercups on cow's udders it would increase their milk yield. They were used to treat burns and sores, as well as rheumatic pains. One of my favourite wildflowers, Devil's bit scabious, grows in meadowland and grasslands. They get their name from a legend that the Devil bit off their root to make the plant a less efficient herb. However, as the name suggests the Devil was not entirely successful because the plant was used in the Middle Ages to treat skin complaints. Another plant that can be seen in this habitat is yarrow which I can attest stems bleeding (just wrap the leaf around the affected area). It is an antiseptic and prevents infection. It is often found on old battlefields. Yarrow might derive its name from an Angle Saxon word meaning healer. If so it is a good name for the plant because it is a cure-all, used to treat kidney, heart, skin and stomach complaints, as well as fevers and flu, rheumatism, and toothache, as well as depression. It is also a good insect repellent. Middle-aged men take note, it can also allegedly prevent hair loss.

Shepherd's purse, so called because its heart shaped seedpods resemble drawstring purses commonly worn by Medieval peasants, produce between two thousand and three thousand seeds per year, which can lie in the soil for a long time, germinating only when the soil is germinated. This flower is also said to stem bleeding (especially after childbirth), as well as healing wounds, and repelling insects.

Yellow- rattle is an important species in meadows. It is semi-parasitic, tapping into nearby grassroots to get some of its nutrients, and therefore it reduces the growth of grass, enabling other wildflowers to grow.

In late Spring, early Summer several of the fields near where I live turn red because of a very well-known and brightly coloured flower, the poppy. Poppies have been used as an analgesic since at least the time of the Ancient Egyptians, where it symbolised the God of Death. However, it is not possible to obtain opium from the red poppies you see in the United Kingdom. In the 15th century poppy petals were seeped in hot water and alcohol in order to produce red ink.

Most people probably associate poppies with the First World War, and on every Remembrance Day, since 1921, many British people wear a paper poppy as an act of remembrance, unlike the French who wear a cornflower and the Belgians who wear a daisy. The reason why we have a poppy as an emblem of remembrance dates to just after World War One. A French lady called Madam Guerin set up an organisation to help the widows and orphans left behind by the war. She was a member of the America and French Children's League and attended a conference in America. At the same conference was an American professor and humanitarian called Moira Michael who, inspired by John McCrae's famous poem, In Flanders Field, had bought paper poppies to give to her friends. Anna Guerin was inspired by Michael and when she got back to France she started making paper and silk poppies and sold them on behalf of the widows and orphans. She later travelled the world persuading people, including

The Wonders of the Wild Places

Earl Haig, to adopt it as the official symbol of remembrance. In 1922 Major George Howson set up the Poppy Factory in Richmond-on-Thames and employed disabled ex-servicemen to make paper poppies. The design of the poppy allows it to be made one-handed, helping one-armed veterans who were restricted in the employment they could do. English and Welsh poppies are still made there. The factory makes thirty-six million paper poppies a year which are then sold on behalf of the British Legion. [37]

There is also a yellow poppy, known as the Welsh poppy, which prefers rockier habitats. It is not related to the better- known red poppies, in fact, its closest relatives live in the Himalayas.

Farming has changed a lot since the 1940s. Increased mechanism and technology has resulted in over intensive farming which combined with the increased use of artificial fertilizers and pesticides has had a severe impact on the quality of the soil. It costs the U.K government between £900m and £1.4bn per year to try and combat erosion and a lot more money to deal with flooding, which is partly caused by poor soil quality. Another consequence of the overuse of fertilizers and pesticides is that food is now less nutritious than it was pre- World War Two. As Isabella Tree says in her fascinating book *Wilding*, "We need to eat eight oranges to get the same amount of Vitamin A our grandparents got from one orange."

However, help is on hand from what Charles Darwin described as *nature's ploughs* and Aristotle described as *the intestines of the earth*- earthworms. Darwin is obviously better known for his theory of evolution, but what many people do not know about him is that he was obsessed with earthworms and conducted many experiments at his home at Down House in Kent. His book on worms, with the snappy title, *The Formation of Vegetable Mould through the Action of Worms, with observations of their habits,* amazingly sold six thousand copies, far more than *The Origin of the Species*. He was one of the first people to realise how important earthworms are in creating fertile, good quality soil.

There are approximately twenty-nine species of earthworm in Britain, and they all play a different role in the soil cycle. These worms fall into

---

[37] Scottish poppies have 4 petals and no leaf and are made at the Lady Haig Factory in Edinburgh.

three categories – anecic, endogeic and epigeic. Anecic earthworms are the most common earthworms found in the U.K. They are reddish brown and make vertical burrows in the soil, leaving casts on the surface of the earth. Endogeic worms are paler and have horizontal burrows and epigeic worms live on the surface, eating compost.

Earthworms improve soil quality by consuming their own body weight of soil per day and ingesting decomposing organic matter, pulling it underground with little hooks on their bodies. The soil they secrete has 10% more nutrients than the soil they ingested. They improve the structure of the soil which results in better drainage, and less flooding. They secrete a substance which sets their tunnels, allowing air to circulate. Earthworms also contain bacteria – up to fifty species of which have been found in the gut of a common earthworm- which, when combined with mucus and digested leaf litter provides rich nutrients, which help plants to grow. It blows my mind to think that, as Isabella Tree says in her book *Wilding*, "In a handful of soil there are more organisms that the total number of human beings that have ever lived on earth." Science is only beginning to understand these organisms, but it is known that not all of these organisms are favourable for plants or for human health. However, earthworms have an amazing ability to select the kinds of bacteria which prevail in the soil, killing off harmful bacteria thus allowing beneficial bacteria to reproduce rapidly.

It is hard to believe that earthworms actually share 70% of their genes with humans. They do not have eyes but do have light receptors. They also do not have lungs, instead breathing through their skin. They are hermaphrodites meaning that they are both male and female although they still require another worm to reproduce with. Contrary to the widely held belief only the head end of the worm can regenerate if it is chopped in half.

Like fungi, earthworms have even been found to be able to decontaminate sites because they can metabolise harmful chemicals and detoxify the soil.

Not everything about earthworms is straightforward though. Scientists have discovered what they call the *earthworm dilemma*. This refers to the fact that whilst earthworms contribute to climate change by breaking down leaf litter and thereby increasing the carbon dioxide emissions from the soil by 33%, they also help sequester carbon because it becomes stored in their faeces. Preliminary results by scientists in the U.S.A. and China have shown that whilst earthworms do emit nitrogen dioxide and aid the emission of carbon dioxide, they might sequester more carbon than they emit.

Farmland is also an important habitat for many birds. According to the R.S.P.B, 28 species depend on a farmland habitat during the breeding season, although many of our feathered winter visitors are also reliant on farmland. The R.S.P.B. and B.T.O. have focused on nineteen of these birds in order to determine whether farmland birds are struggling or doing well. Overall, the picture is not good. Since the 1970s these birds have declined by an average of 48%. However, as always, this is not the full picture. Some birds, such as tree sparrows (which have declined by 94%), corn buntings (-90%) and turtle doves (-98%) have done very badly, mainly due to loss of habitat and food, whereas other birds, such as corvids, birds of prey, and wood pigeons (up 125%) have fared much better.

One bird which is struggling because it is being affected by a disease which kills finches, is one of my favourite birds, the brightly coloured goldfinch. The Anglo-Saxons called goldfinches *thisteltuige* meaning thistle-tweakers and they can often be seen feeding on thistle seeds in the autumn. Their beaks are sharper than other finches enabling them to extract seeds from thistles and teazels. In Victorian times there was a craze for keeping songbirds in cages and wild birds were caught and sold to wealthy urban clients. In 1860 alone 132,000 goldfinches were caught and imprisoned. Goldfinches are sociable birds and can often be seen in flocks. A group of goldfinches is known as a charm from the Old English word *c'irm* which describes their song.

Another bird which you might see on farmland and in gardens is another of my favourites, the starling. [38] These beautiful iridescent birds can be seen moving in large flocks known as murmurations, especially in Autumn. These murmurations can involve thousands, if not hundreds

---

38 According to the RSPB they have declined by more than 80% because of loss of habitat, shortage of food and use of pesticides

of thousands, of starlings. It is not known why starlings perform these breathtakingly beautiful complex aerial displays, like a kaleidoscope, forming different patterns in the sky, but it is thought that it is harder for birds of prey to catch one if they fly in a large group, and that it is warmer for the starlings to fly together. It is also thought that in flying in large groups the starlings can pool information about sources of food. They forage over a twenty-mile area. The best time to see a murmuration is at dusk when they return to their roosts. [39]

    There are eight members of the corvid family in Britain: ravens, rooks, crows (hooded and carrion), jackdaws, magpies, choughs, and jays. As the name suggests carrion crows eat carrion, which is why you can often see them on road journeys, pecking at roadkill, as well as worms, eggs, seeds, and fruit. Hooded crows also eat carrion, as well as invertebrates, eggs, and young birds. Unlike the all- black carrion crow, hooded crows are black, brown, and grey. There are many more carrions then hooded crows, and it has been shown that they do interbreed sometimes so there are hybrids as well. Unlike carrion crows which can be seen throughout the U.K., hooded crows are only seen in Scotland, the Isle of Man, and Northern Ireland. Particularly, magpies, jackdaws, rooks, and crows can often be seen in fields. Rooks and crows can be differentiated by their size, crows are smaller than rooks, and rooks tend to be in groups whereas crows are more solitary. Rooks and jackdaws are both on the increase, up 41% and 135% respectively according to the R.S.P.B. Farmland Indicator. The reason corvids are doing well might be that they are very adaptable, but also might because they are highly intelligent birds. Scientific tests have shown that the rooks can use tools, carefully selecting the right tool for the right task , as well as fashioning tools if none are available ; Magpies are the only birds, apart from pigeons, who can recognise themselves in mirrors, and they can also remember human faces; Ravens can pull strings to receive food but also tactically deceive other ravens so they do not get the food; Jays are able to plan ahead and, like the other corvids, can use objects to raise water levels; Crows can distinguish between familiar and non-familiar human voices. Jackdaws recognise human faces and follow

---

[39] Look on www.starlingsintheuk.co.uk website to find out the best places to see murmurations near you.

human cues, such as being able to track where humans are gazing. Only a few other animals, including chimpanzees, can do this. In short if degrees were awarded to birds, corvids would be getting firsts every time.

As well as being intelligent, crows can be deceptive, building fake nests to fool predators. They also crush ants and rub them over them probably because formic acid wards off parasites.

In rural areas it is thought that rooks can predict the weather. If their rookeries are built high in the trees it is a sign of a good weather, whereas if the rookeries are lower down the summer is said to be wet and windy. Just from my observations I think there might be something in this, but I could not find any scientific studies which supported it, although birds are known to be susceptible to air pressure so maybe that has something to do with it.

Many of you are probably familiar with the nursery rhyme which begins, *4 and 20 blackbirds baked in a pie...* Interestingly this probably does not refer to the birds which we now know as blackbirds and are commonly seen in people's gardens. In the Middle Ages fledgling rooks were made into pies as a rural delicacy and it is thought that this is the origin of the rhyme.

Since the mid-15th century, we have had collective nouns for groups of animals or birds. These were recorded in Books of Courtesy, educational handbooks for young aristocrats. Some of these collective nouns refer to a peculiar habit of that animal or bird, such as an exaltation of skylarks, an asylum of cuckoos, or a wisdom (or parliament) of owls. There is a normally more than one collective noun depending on what the birds are doing. For example, geese in flight are known as a skein, which comes from the Old French word *escaigne* which means a hank of yarn folded back on itself to make a V shape, probably because of the shape the geese form in the air; on the water a group of geese are known as a plump of geese, whereas if they are on land a group of geese is known as a gaggle of geese. With corvids some of the collective nouns refer to the sound made by the birds, so there is a clattering of jackdaws and a scold of jays. The collective noun for a group of rooks is a parliament of rooks[40].

---

40 Also a congregation, storytelling or clamour of rooks

This comes from their habit of gathering together in fields, supposedly listening to their fellow rooks making speeches. However why do we say an unkindness of ravens[41] or a murder of crows? The collective noun for ravens comes from the 19th century when there was the unfounded belief that ravens were bad parents. The collective noun for crows has different explanations. One is that crows have always been associated with omens of deaths, maybe because carrion crows and ravens were often seen scavenging on battlefields after a battle, and crows were often seen around gibbets. Another explanation is that crows are very territorial and have been known to kill other crows who stray onto their patch. Crows and ravens were also seen as guides to the other world.

Corvids have never been popular birds, especially with farmers. This is probably because of their tendency to eat crops. An old rhyme said, *"One for the pigeon, one for the crow, one to rot, and one to grow."* Also corvids, especially magpies, are often blamed for the decline in songbirds because, amongst other things, they eat eggs. However, studies have shown that this is unfair. An R.S.P.B and B.T.O survey, which looked at 35 years of records and the breeding habits of magpies and of songbirds, showed that there was no difference in songbird numbers in areas where magpies where high and areas where there were not many magpies. They concluded that the decline in songbirds is because of human interaction, in the form of loss of habitat and use of pesticides, rather than caused by corvids.

Tim Birkhead at Sheffield University also carried out a detailed study of magpies and fifteen species of songbirds, looking at their population densities and breeding success. He discovered that despite magpie numbers increasing by 5% between 1966 and 1986 this had no obvious effect on songbird numbers. The number of songbirds in woodlands actually increased when there were magpies in the area. Like the R.S.P.B he concluded that the decline in songbirds was due to habitat loss rather than magpies. In a later study he also made the point that domestic cats posed far more threat to songbirds than the much-maligned magpies.

Another study, carried out by the University of Exeter also dispelled the belief that magpies are thieves, coveting shiny things. This myth

---

41 Also sometimes a conspiracy of ravens

began about 200 years ago in a French play called La Pie Voleuse which depicted a magpie as a thief. The team at Exeter carried out experiments. Magpies were given the choice of a pile of shiny objects and a pile of objects sprayed matt blue. Magpies only selected a shiny object twice out of 64 tests, only to immediately discard it in favour of a blue object.

Another intelligent bird which you might often see on farmland, as well as in gardens, are woodpigeons. These are also increasing in numbers and are said to be the most numerous bird in Britain. One of the reasons for this is that they can breed throughout the year, and they are also adaptable. It has been known for quite some time that pigeons are intelligent, but in recent times they have astounded scientists with the level of intelligence they have. Like some corvids, pigeons can recognise themselves in mirrors and can also differentiate between human faces and voices. However, they can also recognise different letters of the alphabet.

Pigeons can often be seen bobbing their heads up and down; this is because their eyes are on the side of their heads and they only have monocular vision, unlike humans who use both eyes to form an image, so bobbing up and down helps them to perceive depth. They can see a stationary object much better than a moving object. However, pigeons can see in ultraviolet light as well as the colours that humans can see. Project Sea Hunt has trained pigeons to spot people in trouble in the sea. It has been shown that pigeons can spot people much quicker than human eyes.

Pigeons can fly high, above 6000 ft (1.8km), and fast, normally about 77 mph, although the fastest recorded is 92.5mph. For many years it was wondered how pigeons were able to travel such long distances and then return to an exact location. The experiments were carried out on homing pigeons rather than woodpigeons, but it has been found that all pigeons, as well as other birds, have the same ability. Pigeons use several tools to help them navigate. It has been known for a long time that pigeons, and other birds, use the sun to help them navigate, but scientists wondered how they managed to navigate when the sun was not there. Recent research has shed light on the answer to this puzzle, by revealing that pigeons have small amounts of a magnetised mineral called magnetite in their beaks (remarkably humans have also been found to have the same mineral in their noses). This works like an inbuilt

G.P.S. system. Also, Atticus Pinzon-Rodriguez and his fellow researchers at Lund University in Sweden have discovered that birds, even non-migratory birds, have a protein called Cry4 which is found in the birds' eyes and enables them to see magnetic fields. This maintains constant levels throughout the day and works in different light conditions.

Scientists have pondered for a while how a pigeon worked out where it was in relation to where it wanted to be. In 2013 an American geophysicist called Jon Hagstrum, who works at the U.S. Geological Survey, had a breakthrough in understanding this. He discovered that pigeons could hear very low frequencies, as low as 0.5 Hz, much lower than the sounds which humans can detect. This is known as infrasound. Hagstrum proposed that pigeons used infrasound to hear their way. They follow infrasound frequencies back to their loft, which has a distinctive sound. He believes that this might explain why even experienced pigeons can sometimes get lost in bad weather, because the stormy weather, might affect the frequencies.

Researchers at Jesus College in Oxford discovered that pigeons also memorise landmarks and use them to navigate their way home. They particularly like linear landmarks, such as roads and railway lines. Each pigeon has a distinct, individual internal map which it can accurately remember on subsequent journeys. Pigeons use these maps in conjunction with their in-built compass.

Pigeons are related to doves. There are three species of doves in the U.K.- collared, stock, and turtle doves. The latter are the smallest species of dove in Britain and are half the size of woodpigeons. Unlike woodpigeons turtle doves, named after the turrr-turrr turr sound they make, are the U.K.'s fastest declining bird and are on the brink of extinction in Britain. Their numbers have declined by 98% in the last 50 years.

They are a long-distance migratory dove in Europe. These birds make a long annual migration every Summer, flying 3107 miles (5000km) from Sub-Saharan Africa. They fly mainly at night, covering 435 miles (700km) in one flight often at speeds of 43 mph. Unfortunately, many are shot as they pass over countries bordering the Mediterranean. However, whilst this has affected numbers the main reason they are doing badly is because of loss of habitat and food. Turtle doves like arable and mixed farmland, especially liking a plant called fumitory, which makes up about 50% of their diet. However, intensive farming and the use of weedkillers, has led to a reduction in this plant.

However, there has been a tiny bit of hope for these beautiful doves in recent years. At the Knepp estate in Sussex a rewilding project has led to a plethora of rare species, including turtle doves.[42] Near where I live in North Yorkshire there is the North Yorkshire Turtle Dove project. There were thought to be fewer than a hundred turtle doves in the whole of Yorkshire, so the North York Moors National Park is working with local farmers and landowners, advising them of how to provide habitats suitable for turtle doves and other wildlife. Regular surveys have shown an increased number of turtle doves have been spotted in the area. It is nice to think that in the future I might get to see one of these birds near where I live.

Many other birds can be spotted on farmland, although all in decreasing numbers, including lapwings, curlews, linnets, and yellowhammers. Linnets get their name because they like to eat flax, a blue flower crop used to make linen. The brightly coloured yellowhammers, which I saw in huge flocks during my childhood, can be identified by their call, which sounds like a little bit of bread and no cheese. They are less abundant in the North and West of Britain and are largely absent from upland areas.

In the farmland around where I live, I regularly see two gamebirds, red- legged partridges and pheasants. The former it is also known as the French partridge because it was introduced into Britain from France in the 18th century as a new gamebird. The latter was brought to Europe by the Romans, but probably brought to Britain by the Normans. Remarkably, thirty species of pheasant have been released at different times in the U.K. so the ones you see today are a hybrid. Pheasant numbers had reduced dramatically, and they were locally extinct by the 19th century, but then they became a popular gamebird and are now widespread.

One of the Autumn/Winter visitors which you might see on farmland is the thrush- like fieldfare which gets its name from the Anglo-Saxon word *felde-fare*, meaning *the traveller over the fields*.

---

[42] For further details look at www.kneppestate.co.uk or read Isabella Tree's brilliant book Wilding .

The Wonders of the Wild Places

It is not only farmland which birds rely on for habitat, but also farm buildings. Barn owls and swallows have both declined in recent years as barns are converted into dwellings. Every farm used to have a barn owl nesting, but now only 1 in 75 do.

Owls are associated with wisdom and intelligence, mainly due to their association with the Greek Goddess of Wisdom, Athene. In his wonderful Winnie-the-Pooh books A.A. Milne reinforced the myth that owls are wise by making Owl the Hundred Acre Wood's Oracle. However, unlike corvids and pigeons they are not particularly intelligent. They have many physical adaptions which enable them to be silent and efficient hunters. Though their large wings and lightweight body enable them to fly slowly and quietly. In addition, they have soft feathers which are covered with a thin hair-like structure which traps air within the feathers' surface, helping a smooth airflow across the wings, thereby making the barn owl less likely to stall when flying at very low airspeeds. The barn owl's foremost wing feather has a row of tiny hooks which deaden the sound of the air hitting the wing enabling barn owls to hear the tiny noises made by the voles, shrews, and mice on which they feed, and allowing the owl to approach its prey undetected. However, their feathers are not particularly waterproof, and they avoid hunting when it is raining because wet feathers make more noise and makes them less efficient at hunters. Thereby, barn owl numbers tend to drop in wet years.

The barn owl's heart shaped face allows them to collect and direct sounds to their inner ears. Their ear openings are asymmetrical which allows them to detect the precise location of the sound. Their hearing is particularly adapted to be able to hear the high pitch frequencies made by the small rodents they feed on, and experiments have shown that they can use this fantastic hearing to catch prey even in complete darkness.

Barn owls' eyes are twice as light sensitive as human eyes, and their low-light vision is highly movement sensitive so if their prey moves they can detect it almost instantly, although if the rodent stays still it will be safe. Experiments have shown that barn owls are not affected by artificial light and even use it to help them hunt, but they can sometimes be temporarily visually impaired if they have sudden exposure to a bright light, such as car's headlights. Their dark eyes also work well in bright sunlight.

The Wonders of the Wild Places

I have a lot of barn owls near where I live and I sometimes see them flying low over rough grassland, quartering the field, listening, and looking for prey. Their white plumage makes their silhouette difficult to detect by the small rodents they feed on.

The sound of a barn owl is very different to a tawny owl. I hear both at night, the twit and twhoo of a pair of tawnies, and the screech of a barn owl as it flies over my back garden.

Growing up I was told by my parents that seeing a barn owl during the day was bad luck. These owls were associated with witches, ghosts, and imminent death during the Middle Ages. However, I have to admit that they are one of my favourite birds, and I love to see them flying alongside hedgerows or perched on fence posts near where I live.

Owls are affected by climate change, wind turbines, road traffic accidents, fluctuating amounts of voles, loss of habitat, and the use of rat poison. A post-mortem of barn owls showed that 91% of them contained rat poison although, in most case, this was not the cause of death. [43]

Swallows are another of my favourite birds, and Spring does not officially start for me until I see the first swallow, normally near my birthday in mid-April. Like turtle doves they also make an incredible migration every year, migrating 6000 miles (9656km) from Southern Africa to Great Britain. Their route takes over the Sahara and the rainforests of the Congo, then via France and Spain, flying low during the day. They travel a staggering 200 miles (322km) per day at 20-35 mph. It is an epic journey and they do not all make it. The weather can affect them. It also affects them during the breeding season in the U.K., although they need some rain in order that they can build their nests our of mud. It takes both birds 1200 journeys to collect the mud to make their nest. No wonder they tend to reuse it for years. However too much rain means there are less insects and that means their chicks do not survive. Swallows mainly eat bluebottles and horseflies. Climate change is having a negative effect on swallows, causing them to lay fewer eggs.

Male and female swallows look remarkably similar although the male

---

[43] If you would like to find out more about how you can help barn owls have a look on the Barn Owl Trust website, www.barnowltrust.org.uk.

has a longer tail. The longer the tail the more attractive the male. A long tail helps the swallow to dart acrobatically. Combined with their slender, streamlined body and pointed wings they are perfectly designed to hunt insects on the wing.

Last Summer I saw probably several tens of swallows and their cousins, house martens, probably up to a hundred of them, diving, and twirling, chirping, and darting over a field. The sun had come out after a period of rain and a cloud of insects hovered over the long grass and wildflowers. It was a beautiful sight. I got into a conversation with someone about how to tell house martins[44], swallows and swifts apart. Swallows have red faces and long forked tailed, house martins are similar but blue-black with a white underside. They tend to fly higher than swallows. Both species like to nest under the eaves of buildings. House martins prefer living in towns and villages; whereas although you do often see swallows in these areas during the Spring and Summer you also see them in more rural areas too. Like swallows, house martins winter in Africa, but very little research has been done on them and although it is thought they probably nest in the Congo rainforests no-one knows that for certain. Unfortunately, due to a reduction in insects (which is caused by an increased use of insecticides), loss of habitat, and adverse weather (a wet spring leads to many house martin fatalities), the number of house martins breeding in Britain has fallen by a staggering 66% since the late 1960s.

Swallows and house martins are part of the Hirundinidae family, but a bird which is like them in many ways, the swift, is not related to them. Swifts have similar shaped bodies, but they are black-brown all over with white chins and shorter, broader wings. They live in cavities of buildings and can often be seen flying high and shrieking in urban areas. I can see, and hear, several of them over my garden on Summer days. They are only in the U.K. between May and early August. These are very mysterious birds because they live in the air at high altitudes. The highest swifts have been observed is 18,700 ft (5700m) over the Himalayas. Young swifts spend the first two or three years of their lives without ever touching the ground. During the First World War a French pilot, flying at 10,000 feet,

---

44 Known as House or window swallows in the rest of Europe.

was stunned to see a flock of swifts, supposedly motionless in the air. They were asleep. As well as sleeping in the air, they also eat and mate on the wing.

Swifts have one of the longest migration routes in the world – 14,000 miles (22,000km) per year between Equatorial and Southern Africa, and the U.K. This is particularly amazing considering these birds only weigh the same as a Cadbury's crème egg. They are long living birds. The oldest known swift was eighteen years old. It was calculated that he had flown an astounding four million miles during his lifetime – that is the same as to the moon and back eight times.

Swifts are adapted for life on the wing with a streamlined body, long narrow wings, and small feet. Their deep-set eyes have moveable bristles in front of them that act like sunglasses, protecting their eyes against the glare of the sun. They eat more species of animals that any other 'British' bird- more than 300 species of insects and spiders.

Swifts have two unique features. Young swifts can last two days without food, in a semi-torpid state, and then instantly regain their weight, making them more likely to survive than other young birds. Swifts cannot feed in wet weather, but are able to fly around storms.

Like house martins and swallows, swifts have also declined, by more than 50% since 1995. The reason for this is that the increased use of insecticides has led to fewer insects, and swifts like to nest in pre- 1940s buildings and many of these have been converted or demolished in recent years.

Whilst swifts and house martins have an air of mystery, swallows have lived alongside humans for millennia, and have many myths and legends attached to them. In Ancient Greece they were associated with Aphrodite, the Goddess of Love. They were seen as messengers, accompanying souls to the next world. Swallows are also associated with Spring and revival and were thought to bring good luck. Traditionally, sailors had tattoos of swallows which denoted the number of nautical miles they had sailed. Each swallow tattoo equated to 5000 nautical miles, so the more swallow tattoos the sailor had, the more experienced mariner he was.

One of the species of birds which particularly like grasslands are green woodpecker. They can often be seen pecking at ant hills because 80% of their diet consists of ants. This is the largest of the three woodpecker species which we have in Britain, but unlike greater and lesser spotted woodpeckers they have much weaker bills and can only chisel soft wood, so rarely drum on trees to communicate. Instead, they make a loud laughing call which sounds like *yaffle*, which gives them one of their country names, familiar to those of us who grew up watching Professor Yaffle in *Bagpuss*. They have a long tongue to extract the ants from anthills. When not in use they can tuck the tongue over their skulls.

Grassland and meadows are also an important habitats for insects including pollinators and moths and butterflies, which I will write about more in the Urban Jungle chapter. The soundtrack to my childhood summers featured the sounds of two insects found in these habitats- grasshoppers and crickets. There are actually 23 species of crickets in the U.K. and eleven species of grasshoppers.[45] They tend to be heard before they are seen. Crickets and grasshoppers both use sound to attract a mate, with each species having a unique 'song.' They make a noise by rubbing their legs against their wings. This is called stridulation. If you hear the familiar chirping noise during the day then they will be grasshoppers, whereas crickets are crepuscular so are more likely to be heard at   dusk. Grasshoppers are smaller and have shorter antennae than crickets. Most grasshopper can fly as well as jump; the exception are meadow grasshoppers who have stunted back wings so cannot fly. Grasshoppers' ears are at the base of their abdomens, whereas crickets hear through sensors on their front legs. Grasshoppers are gymnasts of the insect world, able to jump twenty times their own body length. They can do that because their back legs act like a catapult and their knees act like springs. In contrast to crickets and grasshoppers, their cousins, groundhoppers, tend to prefer shorter grass and like damper habitats, like near ponds and streams.

Grasshoppers and crickets are not the only insects which can jump. There are about 250 species of springtails in Britain. These tiny terrestrial jumping shrimps have lived on earth for about 400 million years and use a hydraulic piston on their underside to jump.

However, the world's greatest jumper, which can also be seen in

---

45 There is more information and useful photographs which help with identification on the British Naturalists' Association at www.bna-naturalists.org.

meadowland, is the common frog-hopper, which despite being only 5-7 mm long, can jump 70cm. This is the equivalent of a human being able to jump over the Great Pyramid of Giza. In order to do this the common frog-hopper uses its very powerful hind legs. It requires a lot of G-Force to get off the ground - 4000 gravities. In comparison a space rocket when it takes off uses a G-Force of five gravities. Common frog-hoppers also have another special talent. They blow bubbles from their bottom and create 'cuckoo spit' which you can often see on grass. This helps keep these amazing insects moist, but also protects them from predators.

Why are meadows, grassland, and farmland important?

These habitats provide shelter and food for a large number of diverse species. If managed properly they also help prevent flooding and store carbon.

An incredible 82% of all the carbon in the terrestrial biosphere is in the soil. The world's soils hold more organic matter than all the other vegetation on earth. Organically farmed land leads to more mycorrhizae and worms in the soil which provides better quality soil, but is also crucial at tackling, and maybe even reversing, climate change. Mycorrhizae are the complex underground network of fungi. This produces a glycoprotein called glomalin which helps transport water and nutrients to the symbiotic plants and stops the fungi from decomposing. However, it also acts as what Isabella Tree in her fascinating book, *Wilding*, describes as, "the superglue of the soil," because it attracts particles of soil, sand, clay, and minerals and produces aggregates, thereby allowing water, air, and nutrients to infiltrate the spaces in between. Glomalin can stay in the soil for up to forty years, protecting organic carbon from being decayed by soil microbes. Therefore, the more mycorrhizae in the soil, the more glomalin, and the better the soil is at storing large amount of carbon. Better still the more carbon in the soil, the more glomalin is produced. Isabella Tree claims that if the world's farmlands were better managed they could capture a staggering ten billion tonnes of carbon per year. She argues that a modest increase in organic matter, of just 0.4% per year, through the restoration of agricultural land, would halt the increase of carbon dioxide in the atmosphere.

Similarly, a Zimbabwean ecologist called Alan Savory, asserts that if five billion hectares of degraded grasslands were restored, ten more gigatonnes of excess carbon could be stored annually which, he claims, would lower dangerous greenhouse gases to pre-industrial levels within a few decades. It remains to be seen whether this is true, but I would argue that it is certainly worth trying. The world's politicians have spent so many decades procrastinating about what to do about climate change that humans are running out of options if they hope to avert catastrophe.

Threats

Over intensive farming since the Second World War has had a huge impact on British nature. Using more land to grow crops, the overuse of fertilizers and pesticides, and the conversion of farm buildings into dwellings has affected many species of birds, animals, plants, and insects as well as damaging the soil. Poor soil quality not only affects how things grow, but also leads to erosion and increased risk of flooding and causes more carbon to be emitted into the atmosphere. Things have got so bad that an article in Farmer's Weekly in 2014 stated that there were probably only one hundred harvests left in the United Kingdom.

As I have already explained worms are also extremely important for good soil quality. Unfortunately, the over reliance of synthetic pesticides has effected the health and mortality of earthworms. Temperature and rainfall also affect earthworm numbers so climate change might have serious implications not only on 'nature's ploughs', but also on soil quality and carbon storage. The #60minworms project surveyed 1318 hectares in England in Spring 2018. The farmers who took part in the study dug small pits across a field and counted the number of worms they saw. The project revealed that in 40% of fields worms were rare or absent. The average field only had nine worms, whereas top fields have three times this number. Only 10% of the fields surveyed have a high number of earthworms. There was a good presence of endogeic worms in 21% of fields, but only 16% of fields had high levels of the vertical burrowing anecic worms, which are important to aid water infiltration and reduce the risk of flooding. The shortage of worms was concerning because as Dr Jackie Stroud from the Rothamsted Research Institute said, "Earthworms influence carbon cycling, water infiltration... greenhouse gas emissions, plant productivity, the breeding success of birds, and even the susceptibility of plants to insect attacks." However, something good did come as a result of the project. 57% of the farmers who took part were so shocked by the low levels, or absence of, worms on their land that after the project they changed their soil management practices. This should help the number of worms increase, but that takes a long time so it might be at least a decade before we know whether it has made a difference.

Over intensive farming has had a severe impact on farmland birds. The R.S.P.B think there are numerous reasons for this including the loss of habitat, less biodiversity, increased use of pesticides, a change in grassland management, increased field sizes, increased numbers of predators, and the weather.

What is being done?

There is cautious optimism about the new Agriculture Act which came into force in November 2020. Unlike the old European Union schemes, the new Act introduces Environmental Land Management Schemes (E.L.M.s) which works on the principle of "public money for public good," and will reward farmers for improving air and water quality, helping wildlife, improving soil health, reducing flooding, and tackling the effects of climate change. However, conservation charities are concerned about the lack of detail and funding and that the E.L.M.s will not come into force until 2027. Time will tell whether this change of direction in agricultural policy will have a positive effect on wildlife which relies on farmland, but in the meantime charities such as the R.S.P.B, Wildlife Trusts and National Trust are providing advice and incentives for farmers to farm in a less intensive and more nature- friendly way.

Some farmers and landowners are already doing things to help nature. In East Anglia there is an innovative scheme to try and increase the number of rabbits. Farmers and landowners are leaving piles of branches near the entrances to warrens, to protect the rabbits against predators and give female rabbits somewhere safe to give birth to their kits. Rabbits live in matriarchal family groups. The dominant female sometimes prevents other rabbits from having kits. It is thought that there is a concern that too many kits in a burrow will attract predators. This new scheme allows subordinate female rabbits space to safely have their kits. It is a four- year project but is already showing signs of success.

Regarding air pollution the government produced a Clean Air Strategy in 2019, but it has been criticised for only concentrating on pig and poultry farms and ignoring beef and dairy farms which are responsible for 40% of ammonia pollution. The National Farmers Union allege that farmers want to change but need financial help to do so.

In 2012, Charles, the then Prince of Wales, in conjunction with the charity Plantlife, celebrated the Queen's Diamond Jubilee by planting a new meadow in every county. Ninety new meadows, spanning an area of at thousand acres, were created. This is a good start, but there is much more to do. Plantlife also have a great campaign called Save our Magnificent Meadows. It is calling for legal protection of the remaining meadows, a national inventory, and the restoration of 120,000 hectares of grassland by 2043 by using natural seeding techniques. Since 2013 Plantlife has created 12,000 acres of new meadows.

Another method advocated for restoring habitats on farmland is the idea of rewilding. This can be demonstrated by the Knepp estate in Sussex who had unproductive agricultural land, lacking in diversity with poor soil quality. Over a period of a few years the land was allowed to return to nature and now the estate has some of the rarest wildlife in the U.K. However it is currently under threat from proposed development.

Another example of a successful rewilding project is the Monk's Wood Wilderness in Cambridgeshire. Sixty years ago, scientists decided to allow four acres of arable land to rewild. Within about a decade a thorny scrubland had emerged, protecting saplings, including many oak saplings planted by jays. Today the site is a, "...complex woodland with multiple layers of tree and shrub vegetation..." and is home to rare species, including the marsh tit and purple hairstreak butterfly.

In East Anglia the Wild East project is a farmer-led rewilding project, which asks farmers to farm in straight lines and allow the edges of fields to be wild. Within its first year eighty landowners have pledged to take part in the scheme and schools, councils and churches are also getting involved, as are the local rail company who has pledged to allow wild areas near all the stations in the region. East Anglia is currently the most intensively farmed part of the United Kingdom so it is hoped that this rewilding project will make a big difference in lowering carbon emissions. Wild Ken Hill in Norfolk is also a big rewilding project, covering 800 hectares of farmland and 200 hectares of freshwater marshland. More than 2500 species have been spotted in the area, including' internationally significant' arable plants, including corn marigolds and stinking chamomile.

What you can do

If you are a farmer or landowner and are interested in creating or restoring meadows or grass land on your land, there is lots of information on the Save our Magnificent Meadows website at www.magnificentmeadows.org.uk. Details of the R.S.P.B, Wildlife Trusts and National Trust projects are on their websites. They also look for volunteers to help survey plants and birds to help monitor whether the schemes are working. One of the best things which people can do is to farm organically and stop using synthetic pesticides. If you have land, you could allow some of it to rewild or just leave wild strips around the edges of the field. Put up barn owl boxes.

Up High and Heaths

Most upland habitats in Britain are in South-West and Northern England, and Wales, Scotland, and Northern Ireland. These habitats can be divided into montane, such as in the Cairngorms in Scotland, moorland, and heathland. Heathland and moorland are remarkably similar habitats with acidic, sandy, or peaty soil, characterised by dwarf shrubs. Moorland has a wetter, colder climate which causes nutrients in the soil to be washed away and few plants and trees to grow, creating blanket acidic bogs, whereas lowland heaths have warmer climates, so more plants and trees grow on heathland. Some plants, such as heather and bilberries, have adapted in these habitats, by reducing their leaves and their height, so they lose less water in the midday sun.

About 70% of the world's heather moorland, like that which covers my local National Park, the North York Moors, is in the U. K. However, between 1946 and 1984 27% of heather moorland was lost. Therefore, there is now less heather moorland than tropical rainforests in the world. This is bad news because moorland provides a particularly important habitat for many species, including many rare ones, such as merlins, red grouse, and golden plovers. Blanket bogs also store 40% of the United Kingdom's terrestrial carbon emissions, and 70% of the nation's drinking water comes from upland areas.

Heather is an evergreen shrub. There are many varieties, but the most common ones are ling, bell, and cross-leaved. It has tiny narrow leaves which prevent it from losing too much water, one of the reasons why it is so important in flood prevention. It grows close together, near the ground, in order that it can survive strong winds. Heather thrives in acidic soil, which is why it grows so well in upland areas. However, it struggles in harsh winters or extremely hot summers.

Other dwarf shrubs, like bog myrtle, crowberry, cowberry, and bilberry also like peaty bogs and moorland. Bog myrtle was once used to make a yellow dye which was used in tanning. Crowberry has dark purple berries which are edible but pretty unpleasant to humans, whereas cowberry has white or pink bell-shaped flowers in short, dense clusters but be careful of the leaves because they are poisonous. Bilberry, known as wimberries in the Welsh Borders and blaeberries in Scotland, has purple-blue berries which can be used to make jam. However, if you live in Scotland or Northern England be careful because there is also bog bilberry, which, apart from the berries, is poisonous.

Many plants cannot grow in acidic soil, but some thrive in it. A good example of this is cotton grass, which is actually a sedge rather than a grass. It has pretty cottonwool- like flowers which was used to stuff pillows in Sussex and Suffolk and used by the Scots to treat wounds during the First World War. Small flowers like lousewort, butterwort, and milkwort thrive in damp acidic soil. The word 'wort' normally denotes that the plant has some herbal qualities, and the former was used to remove lice, butterwort was rubbed on cows' udders to improve their yield, and milkwort was believed to make milk more abundant and used to be given to nursing mothers. Common milkwort, which has alternate leaves, and heath milkwort, which has leaves which are opposite to one another, have a delicate cluster of small flowers which range from red to white to purple-blue depending on whether the soil is acidic or alkaline.

Two other plants which can be seen on uplands and heaths are tormentils (which have four yellow petals) and cinquefoil (which as the name suggests has five yellow petals.) These flowers grow close to the ground. Tormentil is used to treat many conditions including stomach and intestinal problems and mouth infections. Despite being yellow it can be made into a red dye which is still used in artists' paint. Cinquefoil is rich in iron, calcium and magnesium and was used to treat dental issues because it alleviates pain, as well as stemming bleeding. Like tormentil it produces a red dye and was once used to colour cloth and leather.

Another plant you might see is the alpine Lady's mantle which grows in North- West England and Scotland. It is named, like several other plants, after the Virgin Mary, because the leaves were said to be like the folds of her cloak, but its Latin name is more interesting. It is called alchemilla, meaning little chemist, because the water on its leaves was considered the purest water anyone could find, and it was used by alchemists to try and change base metals into gold. It has many herbal uses including as a sedative and diuretic as well as supposedly being good for menstrual problems. It was used to treat broken bones and to aid fertility and facilitate a safe delivery of a child. Dried flowers were placed under a pillow to cure insomnia.

One of my favourite flowers which grows in boggy habitats is the beautifully named and rare grass of Parnassus, which is not actually a grass but a member of the Saxifrage family. It has a very pretty white flower and can be seen between July and October in Scotland, the far North of Wales, Norfolk, and Northern England, especially Cumbria which has it as its county flower (it even appears on Cumbria's coat of arms.) It gets its exotic name from the Ancient Greeks because the cattle on Mount Parnassus liked to eat it.

In the same family as the grass of Parnassus is the starry saxifrage, an Arctic alpine plant with star- like white flowers which is the only alpine plant to like acidic rock and can often be seen growing near upland streams and wet ledges in Snowdonia, Northern Britain, and the coastal areas of Ireland.

Possibly the most interesting plant you can see in boggy conditions is the carnivorous round-leaved sundew. The glistening droplets on its leaves attract midges who think they are water droplets, where they can lay their eggs. Instead, as Charles Darwin was the first to discover, the plant determines whether the object it has caught is organic or non-organic matter. If it discovers that the object it has caught contains nitrogen, hairs curl around the midge, imprisoning it, and hairs at the leaf margins secrete enzymes which digest the hapless midge so it can be absorbed by the plant. The whole process takes a few days after which the leaves open again, and the skeletal remains of the midge are blown away in the wind. Up to two thousand insects, mainly midges, may be consumed by a single plant every summer.

On upland habitats, such as the Lake District fells, in summer there are splashes of pink and white, from clusters of foxgloves. A single flower can produce over a million seeds, so these flowers spread like wildfire. These are also known as witches' thimbles, fairy gloves, and grannies' gloves. Most people are familiar with their cluster of pink, or white flowers, with their spotty interiors. These spots were thought to be the handprints of fairies, whereas the flowers were said to be used by foxes in order that they could stealthily creep up on poultry. These flowers are a fantastic source of nectar and pollen for pollinators, and are beneficial to humans too. Foxgloves are poisonous, but in the 18th century a doctor and amateur botanist called Dr William Withering published *An Account of the Foxglove and some of its Medical Uses with the practical remarks on Dropsy and other Diseases. This* might not seem remarkable but what Dr Withering had discovered was that foxgloves produce two important chemicals, digitoxin and digoxin, two of the most important chemicals used in medicine, which are used today to treat heart disease. However, do not try it yourself because the dose is critical and if too much is given it stops the heart from beating.

A common plant which produces mixed emotions is bracken, which grows in a range of habitats but also likes upland conditions. This is a native British fern, and its name comes from the Old Norse word for fern. On one hand it provides a microclimate and cover for smaller plants, such as bluebells and wood anemones. However, through its underground network of thick, fleshy stems, called rhizomes, it can extend over a large are and increases its cover by 1-3% per year, smothering many other species, such as heather and moorland berries and grasses, contaminating them with chemicals and not allowing them enough light. For this reason it is now viewed by many conservation organisations as a nuisance and a lot of money is spent trying to eradicate it, not easy because it can regenerate even if there is just a tiny bit of it left in the soil.

Another reason why bracken is not popular is that it is dangerous for people and animals. It contains carcinogens, especially one called ptaquiloside which actually changes D.N.A., resulting in a depression of bone marrow, a reduction of white blood cells, and internal haemorrhaging, leading to leukaemia, oesophageal, and stomach cancer in humans and animals. At first it was thought the threat was just from ingestion of bracken. Many animals, especially cattle and sheep, are killed each year because of it. However, there is also a worry that it can be ingested by humans through its spores which it emits into the atmosphere, and by its contamination of drinking water and dairy products. A lot of research is still being done about it, including researchers in Japan who are trying to see if it could actually be used as an anticancer drug, but I would advise people to avoid walking through big patches of bracken where possible. It is also linked to Lyme's disease, which can cause long term health problems, as I know myself because I had it as a child, and forty years later I am still experiencing muscle and joint pain.

It takes a hardy plant or tree to grow on the uplands of Britain. One of these, which is also a valuable habitat and foods source for many species, is gorse, which also grows on heathland. Most people will be familiar with this prickly shrub with bright yellow flowers which smell strangely of coconut, but what many people might not realise is that there are three types of species in the United Kingdom; Western gorse , also known in the North of England by the Norse names furze or whin – the latter features in quite a few place names; dwarf furze, which does not like limestone so avoids my area in the North - East of England, but thrives in Western Britain and Ireland, and ulex minor, which grows in the South- East England, East Anglia, and the Home Counties. The bright flowers cheer me up in the winter because gorse grows pretty much the whole year round.

People in the North of England might be familiar with the Easter tradition of Pace eggs, which goes back many centuries, perhaps even back to the Crusades. Boiled eggs are traditionally dyed, either using yellow gorse petals or red onions, and then rolled down a slope with the winner being the egg which goes the furthest and is still in one piece. There is a similar tradition of jarping where people take it in turns to bash each other's egg, with the first egg to break being the loser. In my family there is a regular pace egg competition at Easter, with some members of the family getting quite competitive.

Gorse was a valuable commodity to people in pre-Industrial Britain because it was used for fuel, feeding cattle, and as a broom. However, traditionally there were very strict rules about harvesting it, with people in Oxfordshire only being allowed to pick as much as they could carry on their backs.

A similar shrub, but without the prickles, is broom. Although these were made to make brooms, that is not how they got their name. They are named after the Old English word *brom* meaning a coarse shrub. Unlike gorse, broom smells of vanilla. Its Latin name is *Planta genista* which historians might recognise as being the name of a family which produced many English kings. The Plantagenets took their name from this shrub and used it as their emblem.

A rarer shrub which can be seen in upland habitats, especially the Lake District, is juniper, which is one of the first trees to colonise Britain after the Ice Age, 10,000 years ago. The name is derived from the Latin word *juniperus* which comes from the words for *younger* and *to appear*. This is because the young berries appear before the older berries have fallen off. It needs very particular conditions in which to grow and grows very slowly. It has declined dramatically in the South of England, where it also used to grow on chalky heathland, but thanks to a successful conservation project by the Plantlife charity, it is making a comeback. This is good news because it provides an important habitat for many species, including rare orchids and butterflies. Forty species of fungi depend on it, and fifty species of insects like this tree.

One of the reasons for the rareness of junipers is that its dioecious, meaning its either male or female, so fragmented habitats have made it harder to reproduce. There is also a pathogen called phytophthora austrocedrae which is killing entire populations of these trees.

It would be sad to lose juniper trees not only because of the habitat they provide, but because gin is made from their berries. Gin has been made since at least the Middle Ages. The name comes from the French word *genievre* or the Dutch word *jenever*, which both mean juniper. The berries need to be dried before being distilled.

The berries have not just been made into gin, they were also used by the Romans as a substitute for black pepper, which was very expensive, and they can still be used in game dishes, bread, and other foodstuffs. The health benefits of the berries are currently being researched to see if they could help diabetics, because the ingestion of the berries encourages the release of insulin. Culpepper also advised juniper berries for improving eyesight (it is said to strengthen the optical nerve), speeding up childbirth (a chemical in the berries stimulates contractions of the uterine, although it was also used to cause abortions), curing skin conditions, improving the memory, as well as helping with coughs and colds, and tummy conditions , and , more bizarrely, snakebites. I do not think many of us need more excuses to drink gin, but next time someone says you are drinking too much you can point out the health benefits.

Juniper was also seen as a deterrent against witches and the Devil and was hung over doorways on the eve of May Day and burnt at Halloween to ward off evil spirits. During times of plague, doctors used to put juniper berries into the beak- like masks they wore because it was erroneously thought that the plague was an airborne disease and that the berries filtered the air.

A more common tree which can be seen in the uplands is the rowan, which derives its name from the Old Norse *raudr,* meaning red, because of its scarlet berries. There are actually two unrelated species in the United Kingdom although it is difficult to tell them apart because their leaves, which look like ash leaves, are very similar. In South-West England it is also known as quickbeam or quickberry, with *quick* meaning life because the tree is associated with life because the berries are full of citric acid and very nutritious for lots of birds and animals. Its flowers are a good source of nectar and pollen for many pollinators. In Wales it was often planted in churchyards to help the dead travel to the next world, and in Scotland and Ireland it was planted near houses to ward off evil spirits and witches. An old Scottish rhyme says, *"Rowan tree and red thread haud the witches all in dread."* This refers to twigs of rowan suspended over the entrance to byres, to protect the livestock from witchcraft. Farming families and their livestock protected themselves with loops made of rowan branches on days on which witches were most likely, namely the shortest and longest days, Equinoxes, and the day before May Day.

Peatlands, which co-exist with heather moorland or upland heathland, are a very important habitat. These are made from partially decomposed vegetation. They are a waterlogged environment which is low in oxygen. This creates peat which grows incredibly slowly – one metre (3ft) per millennia. There are three types of peatlands – blanket bogs, raised bogs and fens. The U.K. has 13% of the world's blanket bogs, with 60% of these being in Scotland. Peatlands are an especially important habitat because they cover only 3% of the earth's surface but store almost a third of the world's soil carbon. In fact, in the United Kingdom more carbon is stored in our peat bogs than in all the trees in Germany, France, and Britain – an incredible 42%.

The main component of these bogs is sphagnum moss. There are at least ten species of sphagnum moss in the U.K. in a multitude of colours. They can hold water the equivalent of twenty times their own weight and can retain water long after the surrounding soil has dried out and survive in floods or droughts. It produces chemicals which make the water more acidic, preventing the decay of dead vegetation, which over the centuries and millennia becomes peat. During the First World War there was a shortage of cotton bandages, so sphagnum moss was used because it is incredibly absorbent and has antiseptic qualities.

Moorlands and peatlands are a special habitat for a variety of birds including some of the rarer birds in the U.K. Black, and red grouse love these habitats. Black grouse have declined by 80% in recent decades, and red grouse has also seen a decline because of loss of habitat. Red Grouse feed on heather shoots, seeds, insects, and berries. They are literally homebirds, travelling very little in their lifetimes. It has a distinctive call which sounds like *Go Back! Go Back! Go Back!* The best time to see black grouse is in April and May when you might see them lekking, which comes from the Swedish word *leka* meaning to play. At the leks between five and thirty black grouse strut, flutter, fight, and make strange noises to hold their place in the lek. The black grouse who dominates the centre of the lek, secures the most females.

Other birds which like these wild, expansive landscapes are curlew. As a child I often heard their eerie onomatopoeic calls when I was walking on the North York Moors, but these are another bird which has declined dramatically in recent decades, declining by a third in England, a half in Scotland, and a whopping 80% in Wales and Northern Ireland between 1995 and 2012 mainly due to loss of habitat. It is not helped by the fact that they lay their eggs on the ground, so they are susceptible to predation.

Another declining bird with a distinctive call is the lapwing, which are beautiful green-purple and white birds. Until the 16th century they were known by their Anglo-Saxon name *hleapwince* meaning moveable crest, because they have a very distinctive crest. Males are the easiest to see, performing incredible aerial displays, tumbling through the air, shrieking *pee-witt, pee-witt.* These displays are thought to be a ploy to distract predators because, like curlew, they lay their eggs on the ground.

The Wonders of the Wild Places

Another trick they have is to modify their behaviour if they are observed by humans by doing things such as visiting a different area to where their nest is. They have seen a huge decline in the last 25 years because of loss of habitat and changes in agricultural practices.

Another bird of my childhood, often heard singing on moorland walks, but also at home on heathland and farmland, is the skylark. These beautiful birds, muses for poets[46] and composers[47] alike, have declined 75% in my lifetime. It makes me sad that fewer people will have the opportunity to listen to their complicated songs, which can last up to an hour. Their songs can contain anything between 160 and 460 syllables. It is thought that larks sing for a variety of reasons including to attract a female, ward off rivals, and deter predators. The Victorians were fascinated with these birds, and they captured larks and imprisoned them in cages, with a particularly good singer costing as much as fifteen schillings. Unfortunately, they also liked to shoot larks and were known to shoot a thousand in a day. However, their recent decline is largely caused by a change of agricultural practices, in particular a change in planting crops from Spring to Autumn thereby losing one of their favourite habitats, Winter stubble.

One of the first migratory birds to return to the British Isles is the wheatear which winters in Central Africa. The whinchat is another Summer visitor from Africa, and the red- bellied stonechat, which is more often seen on heathland and bogs. The latter is named after its call which sounds like two stones being knocked together.

Moorland and montane habitats also support many British raptors (birds of prey) including some of the most common ones and the rarest. A bird which has bucked the trend in my lifetime is the buzzard, which was the brink of extinction in 1900, but thanks to a quadrupling of its numbers in the last fifty years can now be seen in every county of England, as well as in Scotland and Wales, making it Britain's most common bird of prey. This is mainly because they are flexible birds which can live in a variety of habitats, and they also eat carrion so are now found in more urban areas

---

46 It is thought that more poems have been written about the skylark than any other bird.
47 In particular Hoagy Carmichael and Ralph Vaughan- Williams.

too. They can often be seen, soaring in the thermals, mewing, or being mobbed by crows. Thermal currents are updrafts of warm air. By flying in increasing circles within the thermals the buzzard can gain altitude without using a lot of energy.

Another bird of prey which was on the brink of extinction but are now making a comeback is the fastest bird in the world, the peregrine falcon. During the Second World War the Secretary of State for the Air decreed that peregrine falcon nests and eggs should be destroyed because they posed a risk to carrier pigeons who were a vital part of the war effort. After the war, the use of harmful pesticides destroyed many birds of prey populations, making their eggs less viable. Thankfully though since these pesticides were banned and peregrines were legally protected, these remarkable birds have made a comeback and as well as seeing them near cliffs, moors, and mountains they also nest in most of the cathedrals in the U.K. and can often be seen in urban areas.

Most birds are at risk of being predated by this swift and magnificent hunters, but they do particularly like to catch pigeons. I watched one take a pigeon in mid-air and the poor pigeon did not stand a chance. In a dive peregrine falcons can reach the incredible speed of up to 245 mph, making them the fastest thing in the natural world. They are specially adapted to do this. They have cone- shaped bones called baffles near their nostrils which break up the air, allow controlled incoming air, and guide the shockwaves, stopping pressure from damaging their lungs. From studying peregrine's nostrils aeronautical engineers created a metal cone at the opening of a jet engine which prevents the aeroplane from stalling in mid-air. In addition, peregrines have small feathers which pop up when they dive, acting like the flaps on an aeroplane and reducing drag. They are further aided by a third eyelid, called a nictiting membrane, which clears debris from their eyes and lubricates their eyes whilst they are in a dive.

Aircraft designers have been inspired by the peregrine's adaptions. They have designed 'sensory feathers', which are polymer filaments which act like sensors on the body of the aeroplane , providing an early warning system if the engine stalls in mid-air, hopefully allowing the pilot to avoid crashing. This design is influenced by peregrine's feathers which vibrate when they lose airflow. Designers are also looking at incorporating hinged flaps into aeroplanes which would allow them to land at slower speeds and make quick manoeuvres if they needed to. It is hoped that this will reduce noise pollution and allow the aeroplane to fly more efficiently. The inspiration for this is the way that peregrines ruffle their feathers upwards so they can slow down as they engage their prey. Within the next twenty years these adaptions should be a standard part of aircraft design, making aeroplanes safer and more environmentally friendly, and all thanks to this amazing bird of prey.

Peregrines have excellent binocular vision – eight times better than a human, which allows them to see prey from 3km (1.8 miles) away. However, they see better if their head is at an angle, which is why you often see them curve towards their prey. Another reason for the curve is that it helps reduce drag, allowing it to fly faster. Black patches on its face minimise glare so it can focus better on its prey.

Hen harriers are a rarer bird of prey. The genders look different. The females have very distinctive rings on their tail, which is why they are also known as ringtails, whereas the males are pale like, "grey ghosts", according to the Hawk and Owl Trust. These raptors became extinct in Victorian times, but then naturally recolonised the U.K. in the 1950s, before almost becoming extinct this century. In 2011 there were just four pairs recorded in the United Kingdom. This was partly due to persecution and partly because they mainly eat voles and meadow pipits, another moorland bird, which is a favourite snack for birds of prey. Meadow pipit numbers have declined sharply since the 1970s, and vole numbers fluctuate. Organisations such as the RSPB knew that something needed to be done to save these raptors, so they initiated the Hen Harrier LIFE and Skydancer projects which improved habitats, educated landowners and gamekeepers, and protected nesting sites. By working with landowners and gamekeepers to improve moorland management these magnificent birds of prey have been brought back from the brink, and 2020 was a fantastic year for them. Sixty chicks fledged, bringing the total to 141 chicks in three years. There is still a long way to go, but the future does look brighter for hen harriers. I have seen two near where I live in the last 18 months. One flew right across the windscreen of my car, allowing me to see it very clearly. Hopefully, in the next decade or so more people might see them skydancing, which is when they pass food to one another in mid-air.

Visitors to Scotland or Northern Ireland might also see another bird of prey which is now unfortunately extinct in England and Wales, the golden eagle. I regularly saw these huge raptors when I holidayed in the Lake District as a child in the 1980s, but sadly the last English golden eagle died a few years ago.

I have observed someone thinking a buzzard is a golden eagle, but actually the golden eagle is much bigger than a buzzard with a 6ft(2m) wingspan. It is only the second largest raptor in the British Isles though. The largest is the white-tailed eagle which was extinct in the U.K. last century, due to persecution, but in the year, I was born, 1975, Norwegians chicks were released on the Isle of Rhum in the Inner Hebrides, and I was lucky to see them off the coast of Skye about fourteen years ago. More recently they have been released in Norfolk and the South of England and have spread out. Recently one was seen near where I live on the North York Moors.

Maybe a similar scheme to the Skydancer project will happen for golden eagles one day, but for now the best place to see them is the Highlands of Scotland where they are the key predator, eating mountain hares, rabbits, and sometimes foxes, gamebirds, and young deer. They live in nests known as eyries high up in the mountains. These eyries can be passed on from one generation to the next. Pairs of golden eagles cover large ranges looking for food, up to 60 square miles (155 square km) and younger eagles having even bigger territories. Whilst they are not as fast as a peregrine, for a big bird they are very fast, diving at more than 150mph although they soar at a much more leisurely 30mph.

The phrase eagle-eyed has a lot of merit because eagles do have excellent eyesight and can detect movement from a long way away-2km (1.2 miles) for rodents, and double that with regard to rabbits. They see in colour and can rotate their heads 270 degrees, like an owl. Like a peregrine they also have the nictiting membrane (a third eyelid), which helps protect their eyes when they are diving for prey.

Eagles have always been very symbolic of power. The Romans used it on their standards because they thought it symbolised military power because of its associations with Zeus, who they thought of as Jupiter.

The Wonders of the Wild Places

Napoleon, who was influenced by the Romans, also used an eagle as a symbol of power, before in more recent times it being adopted as a symbol by Hitler and the Nazis during the Second World War.

From the largest to the smallest bird of prey, moors are also homes to merlins, which can be best seen in the summer nesting in patches of heather, especially in the Northern Pennines. Incredibly, I have seen two perched next to the A19 in North Yorkshire, near where I live, which just shows you always to keep a look out for things. Habitats are a human concept; wildlife can exist anywhere. Merlins are very small in relation to other birds of prey, only about the size of a blackbird. They are often mistaken for kestrels because they also hover in the air before swooping down onto prey. However, unlike kestrels, merlins prey of choice is other birds, normally meadow pipits.

Merlins were one of the birds of prey used in falconry, which is thought to have started in the Middle and Far East about four thousand years ago but arrived in Europe in 400AD as the Huns invaded Europe from the East. It was at its peak from 500- 1600 AD, before the invention of the shotgun, the breakdown of the feudal system, and new agricultural practices made falconry less relevant. However, during the Medieval period falconry was so popular that literally everyone had a bird of prey. People could not use any old bird of prey though, there were very strict rules in place, and if someone used a bird which did not match their rank in society, they could be punished by having their hands amputated. The Boke of St Albans was written by Dame Juliana Barnes in 1486. She was the Prioress of Sopwell Nunnery. She listed the birds which various ranks of society could own, from the King's gyrfalcon to an Earl's peregrine, to a yeoman's goshawk to a child or servant's kestrel. Ladies were allowed to use merlins. Some of the birds were thought to be more of status symbols rather than of practical use because kestrels tend to eat voles, which were not eaten by humans, whereas a gyrfalcon could catch a rook or heron, two birds which were very popular to eat in Medieval times. Most of the birds were caught in nets along their migration routes in Valkenswaard in the Netherlands, which had a monopoly on the falcon trade until 1937. Falcons were so highly valued that they were literally worth their weight in gold. However, the upside of this was that the first laws to protect birds of prey were passed in England in the medieval period. The penalties were very high. Someone could have their eyes poked out if they stole a falcon.

Thanks to the pen of William Shakespeare a lot of us use everyday words and phrases which are associated with falconry. Haggard originally meant an adult hawk which had been caught from the wild, *to turn tail* meant to fly away, and *hoodwinked* originally referred to the practice of putting a hood over the falcon's head when it was not in use. Under your thumb and *wrapped around your little finger* both relate to how the falconer secured the lead of the falcon to enable them to control it. Bated breath and *end of my tether* both relate to the practice of tethering young, untrained falcons to their perch. A *codger* (or cadger) was an elderly falconer who carried a portable cage called a *cadge* for the falconer. The word which I did not realise was related to falconry is *boozer* which originally referred to a bird of prey which drunk heavily.

Other species of birds also prefer upland habitats including the largest British corvid, the raven. Like all corvids these are extremely intelligent birds. Not only can they imitate humans as well as other birds and animals, but they can also make and use tools. In a logic test ravens had to pull a piece of string to get some food. They all achieved it, some of them in under thirty seconds. They can also be deceptive. In experiments it has been shown that if they know another raven is watching them, they will pretend to hide food in one place whilst hiding it elsewhere. However, even more remarkably, experiments have shown that ravens are capable of communication and empathy. A study in Austria discovered that ravens could indicate an object to another bird by pointing at it with their beaks, and they can also hold up an object to show it to another bird. It was previously thought that only primates did that. Despite being quite solitary birds ravens can recognise another raven after at least three years. Jorg Massen at the University of Vienna studied captive ravens in social groups. Ravens asserted their dominance over low-ranking ravens. The lower ranked raven will normally making a submissive noise but it if does not, confrontation often follows. Other ravens in the group, especially the female ravens, showed signs of stress because they did not want a confrontation. In some cases, it was the low- ranking raven, which was the dominant one, not even primates act that way.

The Wonders of the Wild Places

Like other black birds ravens were associated with death and thought of as a bad omen and cultures from Tibet to Greece thought they were messengers for the Gods. In Norse mythology Odin had two ravens – Hugin (Thought) and Munin (Memory) who flew round the world and reported back what they had seen.

The uplands and heaths are also good places to see rarer birds. if you go hiking in the Cairngorms you might be fortunate to see a ptarmigan. These birds are designed for cold weather. They have feathered legs and feet to keep them warm, and their grey, black, and brown plumage changes to white in the Winter so that they are well camouflaged against the snow.

Ptarmigans are not the only bird or animal which changes their plumage in Winter. In the Cairngorms, mountain hares also have this ability. In the Northern Hemisphere more than twenty other species can change colour depending on the season. There are two ways of doing this. The first are biochromes which are microscopic natural pigments which produce colours chemically. They absorb some colours and reflect others. Different amounts of daylight and a change in temperature triggers a hormonal reaction which causes the animal or bird to produce biochromes. The other way of changing colour are microscopic physical structures which act like prisms, refracting the light. A good example of this, which you will only see in a zoo in this country, is a polar bear. Polar bears actually have black skin but look white because they have translucent hairs. When the light shines on their hairs the hair bends it a bit, causing the light to bounce around, so that some of it makes it back to the surface of the skin and the rest is deflected outwards.

Climate change is affecting the ability to change colour, because there is little or no snow so the bird or animal can easily be spotted by predators. The University of Montana is looking at whether species could adapt evolutionally to these changes quick enough to allow their populations to recover. I hope they can. Mountain hares can delay changing colours by three weeks depending on conditions so maybe they will be able to continue to evolve to be able to hold off changing for longer.

Some mammals also prefer upland habitats. The one which you are most likely to see, especially in Scotland, is Britain's largest land mammal, red deer.[48] Now is arguably the best time to see red deer, as well as the other deer species in the United Kingdom, because there are said to be more deer now then there were at the time of the Norman Conquest a millennium ago. This has an impact on the environment, especially woodland habitats.

Red deer are huge animals, with stags (male red deer) weighing between fourteen and thirty stone (90-190 kg)[49]. Stags have antlers which get bigger as they get older. They shed their antlers every Spring before regrowing them again in late Summer. Growing antlers takes a lot out of the stag and takes a lot of energy, so stags tend to lose weight at the end of the Summer. The antlers are made from bone. They are covered in "velvet" which nourishes the blood vessels of the antlers by providing it with oxygen and nutrients. Once their antlers have grown the deer loses the 'velvet' by rubbing their antlers against a tree.

The best time to see, and to hear, red deer is in the Autumn when mature stags rut to try and win the right to mate with the does (females) in the hareem. At dusk or dawn, you can often see stags charging at each other and then gladiatorially locking antlers and fighting so viciously that sometimes the loser loses his life.

Other deer species which can be seen in Britain are roe deer, fallow deer, and muntjacs, although all these species prefer woodland and farmland. Roe deer are also native to the United Kingdom. They are smaller than red deer, and the bucks (males) have smaller antlers, with a red coat in the Summer and brown-grey one in the Winter and a prominent white rump, which helps with identification. They are more solitary than red deer and I often see one shyly looking at me from behind a tree when I walk in local woodland. Male roes rut earlier than the red deer so look out for them in the Summer, especially at dusk and dawn when you can often see them at the edge of woodland. Fallow deer were originally introduced by the Romans, although the current ones can be traced back to the Normans who kept them in deer parks. They are smaller, about 20 inches(50cm) tall, and can be in a range of colours. Commonly they are tan with white spots and a white rump with a distinctive black horseshoe-shaped border,

---

48 They can also be seen in woodland
49 The average man weighs 12 stone 4/79kg

but they can also be much paler, even white, as well as black. They are the only British deer to have palmate (spreading out like a palm) antlers. They are common in England and Wales but rarer in Scotland. Their hoofprints can be differentiated from roe deer because fallow deer have more elongated feet. They are often seen in large herds. Muntjacs are recent in-comers to the United Kingdom having been brought over to Woburn Abbey in Bedfordshire in the 20th century from China, then escaping and now being widespread across the United Kingdom. I saw one in the Langdales in Cumbria about fifteen years ago and was amazed by how small it was, not much bigger than our dog at the time. They are russet red most of the time except for Winter when they turn grey. They have small antlers. They are quite solitary, like the roes, but like the other species of deer they do like to eat saplings so can cause problems in woodland. In addition to these species, Chinese water deer and sika deer can also be seen in the United Kingdom although, like the muntjacs, these are not native species. Chinese water deer are the only deer species in the United Kingdom which does not grow antlers. Sikas look quite like fallow deer. They were introduced from Asia in the 19th century for wealthy Victorians' deer parks but have escaped and can now be seen in Scotland, Cumbria, Dorset, and the New Forest.

Heathland is one of the rarest habitats in the world, much rarer than rainforests, but 20% of the world's heathlands can be found in the United Kingdom. However, this is much less than we had before the 19th century.[50] In the last century alone half of heaths have been lost. Many of those heaths dated back to at least the Bronze Age. Like moorland, heaths are man-made habitats, and many have reverted to woodland or being converted to agricultural land. There are several different types of heath in Britain: - U.K. dry and humid heath, Eastern Continental heath, South-West Oceanic dry heath, Central warm oceanic heath, upland transitional wet heath, drier forms of wet heath, maritime heath, and montane heath (moorland.) The wetter and montane heaths are more common in Northern and Western Britain, whereas the oceanic heaths prefer lower altitudes. Like moorland these habitats are characterised by dwarf shrubs

---

50 We only have a sixth of the heaths we had pre-1800.

like gorse, heather, bilberries, bog myrtle, and crowberries, as well as plants like sundew and cotton grass. They are all important habitats but maybe the most important one is the Eastern Continental heath which can be seen in the South of England from Kent to Dorset. This heather dominated heath is not particularly biodiverse but is home to many rare species and many of our reptile species.

Two of these rare species are nightjars and nightingales. Nightjars are nocturnal Summer visitors[51]. They fly here from Southern Africa. They are also known as fern-owls or night-churr because the males have a churring call which is an incredible 1900 notes per minute. There are under five thousand nesting pairs in the United Kingdom, but numbers are increasing.

Like skylarks, nightingales have a reputation of having a beautiful song. In fact, the species have 250 songs in their repertoire with a single nightingale having about 180 riffs. A typical aria, which can be heard up to a mile away on a still night, lasts about half an hour but a nightingale was recorded singing continuously for almost a full day. The songs are loud as well as complex, reaching up to 95 decibels. The male nightingale with the loudest and most complex songs are the ones who get the girls. In order to try and attract a female they sing day and night, but night- time is the best time to hear them.

Like the nightjar, the nightingale is a Summer visitor from Africa. The male nightingales arrive from Senegal, the Gambia, and Guinea-Bissau in mid-April after a 3000-mile journey because there are less predators in Europe, so their young have a better chance of fledging. The female nightingales arrive a few weeks later. They fly at night to avoid predators. The United Kingdom is the most northernly point in their range because nightingales are quite fussy. They require the temperature to be between 17-30 degrees Celsius during July and they do not like height, so they are very rarely seen above 600 feet, so North Yorkshire is the most northernly point you can see, or hear, them. The best place to view this iconic birds are in Kent, Sussex, and Suffolk.

Only a couple of centuries ago it was possible to hear nightingales in London, although they had died out by World War Two, so it is extremely

---

51 They are in Britain from late May to mid-August

unlikely that they ever did really sing in Berkley Square. However, Keats noted hearing them on Hampstead Heath, in the early 19th century. Unfortunately, the Victorians obsession for catching and imprisoning songbirds had a hugely negative impact on nightingales. In 1830 a gamekeeper in Middlesex could catch up to 180 nightingales a night, before selling them to people in London for eighteen schillings a bird. As nightingales do not thrive well in activity many of these captured birds died. As a Victorian naturalist, Richard Jefferies, said, "The mortality was pitiable. Three quarters of these creatures who were singing a week before…in the lanes of Surrey, would be flung into the gutters of Seven Dials or Whitechapel."

Together with pesticides, changes to their habitats, and climate change nightingales are in trouble, and the Knepp estate in Sussex and Lodge Hill in Kent, both of which are currently at risk of being lost to development[52], are the only places in the United Kingdom where these birds are flourishing. However, nightingales like untidy hedges and hedgebacks so maybe the proposed new payments to farmers to manage their land so it is good for nature, might benefit the nightingale. I certainly hope so. It would be sad if no- one in the United Kingdom could listen to their beautiful song.

Another rare bird which can be seen in some Southern heaths, mainly along the Southern coast of England, is the Dartford warbler. This is also used to be known as the furze-wren, after the Norse name for gorse, the Dartford warbler's favourite shrub. Unlike the nightjar and nightingale, the Dartford warbler does not migrate and that has been to its detriment because this little bird, which only weighs ten grams, is at the very northernmost of its range in Southern England and does not like harsh winters. The infamous cold winter of 1962/1963 was the final straw for many of these little birds and by the Spring of 1963 there were only a dozen surviving pairs. However, whilst it is still a rare bird, this is a bird which is liking the effects of climate change, especially the warmer winters, and their numbers are increasing. There are currently about three thousand pairs in the United Kingdom.

If you are walking in heathlands, you might also see a hobby. It is

---

52 The RSPB have a Save Lodge Hill campaign if you are interested in getting involved.

another summer visitor so look out for it between April and October. Unlike other birds of prey this raptor eats insects, such as dragonflies, and young swallows and house martins, which it catches on the wing, zigzagging after its prey.

This falcon's Latin name is *Falco Subbuteo*. When Peter Adolph came up with an idea for a table football game just after the Second World War he could not think what to call it. After racking his brains to think of a suitable name for his new game he came up with the idea *hobby*, but this was rejected on the basis that he might as well just call it *game*, so he decided to call it after the hobby's Latin name and named his new game *Subbuteo*. Many children, including my brothers and sister and I, loved his game and he became a millionaire.

Five species of British reptiles can be seen on heathland – the common lizard, sand lizard,[53] adder, slow worm, and smooth snake. Common lizards are brownish grey with rows of darker spots or stripes down their back and sides. Females are paler and have plain bellies. It is found throughout the British Isles except islands such as the Scottish islands, Isle of Scilly, and the Channel Islands. It is also Ireland's only native species of reptile.

Like other British reptiles common lizards are in a dormant state, called brumation, during winter because they are cold- blooded. Therefore, the best time to see them is during the summer when you can spot them warming themselves on a rock.

The common lizard is unusual amongst reptiles in that it is viviparous, which means it incubates its eggs inside its body and gives birth to live young, in common with mammals rather than other reptiles which usually lay eggs and then sit on them to incubate them.

They have a good trick for avoiding predators. They can shed their tails and then regrow a new one.

Slow worms[54] are actually legless lizards rather than worms but look quite snake-like. They are viviparous, like common lizards, but unlike other British reptiles they do not bask in the sun but can often be seen snuggled up in compost heaps and under piles of logs. They have ears, unlike snakes who "hear" through picking up vibrations. Slow worms

---

53 See the deep blue sea chapter
54 These cannot be seen in Ireland.

The Wonders of the Wild Places

spend a lot of time underground, so they have eyelids which they use to blink away dirt from their eyes.

Adders are the only venomous snake in the United Kingdom[55] although there have only been ten fatal adder bites in a century and the last one was about 50 years ago. However, if you have dogs, it is a good idea to keep them on the lead when walking in heathland and moorland, not only to protect ground nesting birds but also so they do not get bitten by an adder. Three dogs in my town died of adder bites last year. [56]

Although adders are occasionally black most of them have a distinctive zigzag pattern, which are unique to each snake, making them easy to identify. They are viviparous and hibernate from October to February. Like common lizards they can often be seen sunbathing in warm weather. They eat small rodents and lizards and are eaten by crows and buzzards.

The rarest reptile in the United Kingdom is the smooth snake. It likes sandy heathland and can be seen at a handful of heaths along the South coast from West Sussex to Devon. Unlike adders, smooth snakes are constrictors, crushing their prey (normally other reptiles, like sand and common lizards, and small mammals) to death. They are eaten by various birds, including raptors, and badgers, foxes, and weasels. They look like an adder but do not have the distinctive zigzag pattern.

Five thousand species of invertebrates also rely on moorland and heathland. Amongst them are the emerald green tiger beetle, and the silver studded blue butterfly. The latter only exists in small colonies in Southern England and Wales. The larvae live on bell and cross-leaved heather, as well as gorse bushes, before emerging as butterflies in June. They can be seen until late August. They have a symbiotic relationship with ants, who protect the caterpillars in return for a sugary substance.

Upland habitats, such as the Lakeland fells, also provide a home to some rare invertebrates including the Northern dart moth, Northern pill beetle

---

55 Slow worms also cannot be seen in Ireland.
56 If you or your pet does get bitten by an adder the NHS advises that you do not try and suck or cut the poison out and do not tie a tourniquet. Instead, keep calm, try to keep still, and go to hospital or the vets as soon as possible. Speed in obtaining an antidote is the key to survival.

and the rove beetle. The Northern dart moth lives mainly on crowberry and has a two-year life cycle. They can only be seen in Northern England and the Scottish mainland and islands.

Rove beetles are the largest extant family of organisms, numbering at least 63,000 species, with a thousand being found in the United Kingdom. They have existed for at least 200 million years. One of the most common rove beetles to be seen in the U.K. is the Devil's coachman beetle, which is also known as the Devil's coach beetle or the Devil's coachman. It has been associated with the Devil since the Middle Ages when it was believed that the Devil assumed the identity of this beetle in order to eat sinners. The reason why it got this reputation might be that it has a scorpion-like habit of squirting a foul-smelling liquid from its abdomen if it is attacked, and it can give a nasty bite. These beetles play an important part in recycling nutrients by eating leaf litter and insects, which they catch by covering in a glue-like substance which sticks to their prey. They use their large jaws and front legs to turn the unfortunate insect into a ball, before chewing it until it becomes liquidised then finally ingesting it. The best place to see them is under stones, and around the margins of freshwater.

Also, in this group are dung beetles. There are sixty British species of them. Dung beetles live on every continent, except Antarctica. Unlike some of the ones in hotter countries, which you might have seen on David Attenborough documentaries, British dung beetle are tunnellers rather than ball rollers, preferring to live inside the pile of poo. These insects play an important role in the decomposition process and, like worms, they are a good indicator of healthy soil. By breaking down the dung, the beetles help improve the soil's ability to retain water and increase the amount of nutrients in the soil, helping plants to grow. However, like many other species, these extremely useful invertebrates are in widespread decline mainly because of the use of worming medication in animals, which are toxic to the dung beetles.

Why are uplands, heathland, and moorland important?

- They provide habitats to some of our rarest species.
- A well-managed upland can help the retention and flow of water, reducing flooding and erosion.
- Well-managed peatlands are important carbon stores.

Threats to uplands, heathland, and moorland
UPLANDS
- Erosion – from water, the weather, and walkers.
- Overgrazing. George Monbiot believes that the Lakeland fells have become "sheep wrecked..." He told ITV news in 2015, "I see an ecological disaster zone ...a place which has lost most of its ecological function...and structure."
- Pollution- from industry and agriculture.
- Climate change. Warmer Summers could mean that grasses and dwarf shrubs grow higher, meaning they might outcompete mosses, lichens, and other plants. It might lead to more erosion and wildfires. Wetter Winters might cause more erosion but could help some plant species. The Arctic species, who are at their most southernly range in Britain, might be the worst affected, with some becoming extinct. Other species might migrate upwards, making species that require upland habitats to survive more vulnerable.
- Bracken not being managed and getting out of control, posing a threat to wildlife, livestock, and people.
- Peat drying out leading to emissions of carbon – currently more emissions than all the oil refineries in Britain.
- Burning heather – it destroys habitats and contributes to carbon emissions.

HEATHS
- Under management- land reverting back to woodland.
- Development

What is being done

There are numerous projects being carried out at the moment to improve moorland, peat bogs, and heathland. Here are some of them: -

EXMOOR MIRES PROJECT
A quarter of Exmoor National Park is moorland which fall into five categories: - coastal heath, northern and southern heather moors, central grass moors, and Brendon heaths. The first of these, coastal heath, is an extremely rare habitat.

Exmoor is wilder and wetter than other moorland in Britain and has the greatest proportion of well drained, mineral rich soil (85%) in all the upland areas in England and Wales. However, the blanket bog and peat that covers the central moorland had dried out after years of drainage and domestic peat cutting. It had lost its ability to store water, thereby meaning the habitat was less diverse than it should be. In 1998 the project started to restore the blanket bog by blocking ditches and improving water management. Between 2010 and 2020 the partnership spent £4.5m restoring 3000 hectares of peatland. This has resulted in an increased water table. The project has improved water quality, carbon storage, and biodiversity which has improved by 31% already.

MOORS FOR THE FUTURE
This partnership project is working on improving peat bogs and moorland in the Peak District National Park. One of the problems they have in the area is that there are exceptionally high levels of heavy metals, such as zinc, copper, and lead, in the peatlands as a result of the smoke from surrounding cities during the Industrial Revolution. These heavy metals are polluting water ways, and acid rain has caused the peat, in some cases, to have the same pH as lemon juice, which has destroyed sphagnum moss, which is so vital for water retention and storage. This has resulted in the peatlands in the area releasing a massive 800 tonnes of carbon into the atmosphere instead of storing carbon. The situation has been made worse by drainage, peat extraction, and heather burning.

Moors for the Future are restoring the peatlands and moorland by blocking drainage and creating natural ways of controlling water flow and retention, replanting sphagnum moss and other species. Part of the project was the Kinder Catchment Project, a five-year project to re-vegetate and restore a blanket bog. Five years after the project was completed the peat is stabilised, the water quality and habitat has improved. Bare peat in this area has been reduced by 75%, leading to more plants, mosses, and lichens in the area. Their ongoing MoorLIFE2020 project is stabilising and revegetating a square kilometre of bare peat, installing 15,000 mini-dams, cutting back over dominant species and planting sphagnum moss and other bog species.

The Moors for the Future project has been ongoing for ten years and has already restored 5000 acres of moorland and 6000 acres of SSSIs, planted 40 million plugs of sphagnum moss, 250,000 moorland plants, and 40,000 new trees.

PENNINE PEAT LIFE
This is a project to restore and protect peatlands and moorlands in the North Pennines, a special habitat for many rare species. Like Moors for the Future, Peat LIFE are covering areas of bare peat with heather brash, which is a mix of heather, cotton grass and other plants, and sphagnum moss, as well as building natural dams to prevent water run-off. They have restored an astonishing 35,000 hectares of peatlands. To put that in perspective that is an area three times the size of Newcastle-Upon-Tyne.

NORTH YORK MOORS
I was brought up near, and still live near, the North York Moors National Park and I love the expansive heather moorland, especially in late Summer when the heather is in bloom. The scent is incredible. The North York Moors are recognised within Europe as being an especially important habitat for many internationally important species including merlins and golden plovers and the rare Duke of Burgundy butterfly. However, like other moorland areas of the United Kingdom these habitats have been deteriorated by lack of management and grazing, burning rotations of the heather (which helps the heather regenerate but affects other species), burning wet peat (which has allowed more carbon to be released into

the atmosphere), erosion (peat burning, water, weather, and hikers) and digging watercourses (thereby drying out bogs.) Between 1948 and 1984, 27% of heather moorland was lost within the National Park. The North York Moors National Park is also concerned that that climate change will lead to more droughts and flooding. The plan to turn things around involves appropriate grazing and burning regimes, blocking drainage to repair peat bogs, controlling bracken and scrubland, and better management of the land.

Currently there are various projects going on in the National Park, including Slow the Flow project where natural methods have been used to reduce the flow of water from the moors and to store excess water, thereby greatly reducing the risk of flooding in the town of Pickering and surrounding areas.

YORKSHIRE PEAT PARTNERSHIP
This project is led by the Yorkshire Wildlife Trust. In order to restore blanket bogs the partnership is filling channels, building dams and planting sphagnum moss and native plants. In its first five months the partnership had already restored just over five thousand hectares of blanket bog and in doing so had saved 48 tonnes of carbon from being released into the atmosphere.

LAKELAND LIVING LANDSCAPES
This is a project to restore and protect the Lake District fells. Over grazing, walkers, and climate change have combined to adversely affect the Lakeland mountains, which include England's highest mountain, Scafell Pike.

The Lake District has several rare species which are at the very edge of their range so any change in climate could lead to local extinction. Weather and walkers, like me, have also eroded the fells, exposing peatlands. Fix the Fells is restoring upland paths in an attempt to reduce erosion. In addition, the Lakeland Living Landscapes project is wanting to improve habitats, to allow species to move through the landscape more easily, and is also stabilising soil and peat, improving water flow, and improving biodiversity. Already by replanting sphagnum moss and other species, blocking drainage, creating natural dams, and removing species like rhododendrons, they are seeing an increase in biodiversity and the restored blanket bogs are now capable of storing large amounts of carbon.

Similar projects are taking place in Snowdonia, the Cairngorms, and many other upland areas. In addition, bracken is being managed and removed where necessary. What I find amazing is how quickly nature returns to these restored habitats. There is still more to do but this gives me great hope for the future. However, these projects are expensive and now that E.U. funding is no longer available the government must ensure that the necessary funds are given to these projects.

From May 2021 it became illegal to burn blanket bogs in protected areas. This is a good start, but the legislation does not go far enough. It will still leave 40% of blanket bogs and 70% of shallower, and normally more degraded, upland peatlands unprotected. In contrast, the Scottish government has banned all burning except in extremely limited circumstances. There are also concerns that it will be difficult to enforce and to monitor.

Heathland is also being restored. There are various projects including the National Trust's project to restore heathland at Cragside in Northumberland, the Forestry Commission's project at Honiton in Devon, and restoration work in the Wye Valley. In Hertfordshire, the Hertfordshire Wildlife Trust is restoring Hertford Heath, one of the last remaining heathlands in the county. Trees and scrub have been cleared and the soil has been improved. The Dorset Wildlife Trust has the Great Heath Project which is restoring 850 hectares of heathland between Dorchester and Ringwood. Conifers and rhododendrons have been removed.

There are various projects to help particular species. The Amphibian and Reptile Conservation Trust's Snakes in the Heather project aims to help the smooth snake by improving habitats and is also hoping to improve habitats in Scotland in order to support adders. The Wildlife Trusts are one organisation who is trying to improve habitats in order to help sand lizards. Projects in Wales have seen an increase in sand lizards. Also in Wales there are plans to reintroduce golden and white-tailed eagles, which have been extinct in Wales for 150 years. The RSPB and other organisations have had several projects in recent years to try and improve habitats, and attitudes, in order to help hen harriers. These projects have been very successful. After years of persecution hen harriers had a record breeding year in 2020, with 84 chicks fledging. This might not seem like a lot but when a species is on the brink of local extinction every chick that fledges, or young that survives and breeds, is important. Part of this success is due to the Swinton Estate in North Yorkshire where Lord Swinton and his gamekeeper, Gary Taylor, are actively managing the 20,000-acre estate to encourage and protect hen harriers. Just on that one estate there were 23 chicks this year. Hopefully their success will encourage other big landowners to manage their land better.

What you can do

- Volunteer on one of the projects (there are further details at the end of the book.)
- If, like me, you like walking on the upland areas stick to paths.
- If you are a landowner, restore peatlands and do not burn moorlands.

Simply messing about in boats

We have many different inland freshwater waterways in the United Kingdom from canals to rivers, lakes to ponds, marshes, and fens. These provide an important habitat for many species. Fens are similar to blanket bogs, but they are rich in nutrients which is food for flora and fauna. Marshes have alkaline or neutral water and are mainly dominated by reeds and rushes. There are 109,000 miles (175,000 km) of streams and 56,000 miles (90,000 km) of rivers (essentially a bigger stream) and just over 1500 miles(2500km) of canals. In additions there are 10,000 lakes and about 470,000 ponds in the United Kingdom. Ponds support 66% of freshwater species in the United Kingdom. More than a hundred priority species are just associated with ponds. As John Lewis Stempel says in his book *Still Water*, "A dip with a net in a pond is like the original lucky dip."

Many birds rely on wetland habitats including herons, kingfishers, and dippers. Herons mainly eat fish, but also eat frogs, ducklings, and small rodents. If they do eat something furry, they dunk it in the water first to make it easier to swallow. Herons live quite socially so look out for heronies (groups of herons) in trees near rivers, tarns, and lakes. Herons were the property of the Crown in the Middle Ages and penalties for poaching them were extremely high, in some cases resulting in an amputation of the hand.

Kingfishers are one of the most colourful birds in the United Kingdom and often you can see a flash of blue as they dart up and down a river, although keep an eye on branches which overhang the river, you might see one perched on it looking for fish. I was lucky enough to see some recently at Nidd Gorge near Knaresborough. As I was eating my picnic one came onto a nearby branch before diving in and catching a fish, which it then bashed on a rock and swallowed whole. It was a wonderful sight.

Kingfishers like slow moving shallow rivers and streams which have an abundance of fish. They can only eat small fish, up to 3inches (8cm) long although they prefer fish which are precisely 1.2 inches (23mm) long. When they see a fish, they dive into the water. They have a third eyelid which comes over their eyes, so they are effectively blindfolded as they catch the fish. The reason they beat the fish on a stone or branch before eating it is to kill the fish, and relax the stickleback's spines, making it easier for the kingfisher to swallow it whole. A kingfisher needs to eat a lot of fish a day- at least its own bodyweight.

You might think that kingfisher is an obvious name for a bird that eats lots of fish but, in fact, kingfishers were originally called *isern* which referred to their beautiful blue plumage. From about 1000, *fisher* appeared but it was not until the early 15th century that they became the King's fisher. The Greek word for kingfisher is *halcyon*[57], as in "halcyon days", which originally referred to calm, fine weather. Dead kingfishers were used as compasses, it was claimed that their beaks faced the prevailing wind. Kingfisher feathers were said to ward off thunder.

Kingfishers were also associated with peace and love because it was said it was the first bird to fly from Noah's Ark after the flood, supposedly getting its orange breast from the sun, and its blue plumage from the sky. The poet, Gerald Manley Hopkins, described kingfishers as *catching fire on a bright day*. Anyone who has seen these beautiful birds fly down a river on a summer's day will know what he meant.

They live in burrows in the riverbank. It takes about a couple of weeks to make. Unfortunately, kingfishers are vulnerable to harsh Winters and pollution. Sadly only 25% of fledglings survive, and only 25% of these will live to breed the following year.

Kingfishers like lowland waterways, if you are walking near an upland river look out for a dipper, which as the name suggests dips up and down. These dark birds, with a distinctive white bib, like fast flowing upland rivers. They nest in natural or man-made crevices, such as waterfalls, caves, weirs, and bridges. The same nesting site is used by several generations with one site being used for 123 years. Dippers live on caddis fly larvae. They submerge themselves in the water to find the larvae. They have a transparent third eyelid which enables them to see underwater.

Moorhens and coots are commonly seen wetland birds. Coots are black with a white spot on their heads, whereas moorhens are black

---

57 From the Greek Goddess Alcyone

with distinctive red beaks. They might look similar, but their habits are quite different. Moorhens spend more time out of the water than coots. Moorhens are omnivorous eating snails, fish, insects, and berries, whereas coots eat weeds and algae. The coot has a white bill whereas moorhens have a red or orange bill with a yellow tip. The colour of their bills indicates how healthy they are, the redder the beak the healthier the bird. If a moorhen has a red beak, it indicates it has low levels of bacterial infection whereas a yellow beak denotes that it is less resistant to disease. Not surprisingly female moorhens prefer the boys with the redder beaks. Moorhens are not particularly good parents and often put their eggs in other moorhen's nests.

Probably the most seen wetland birds are ducks, geese, and swans. There are many species of duck, and they can be found on every continent, except Antarctica. They live in freshwater or seawater and are omnivorous. You often see ducks preening themselves. This not only keeps them clean but also helps to ensure that their feathers stay waterproof. They cover their feathers with a waxy substance. They have a third eyelid which acts like goggles, allowing them to see and swim underwater. The shape of their bills acts like a filter, enabling them to sieve mud, looking for food.

Have you ever seen a duck standing in a river in the middle of Winter and wondered how it can withstand the cold temperature? The answer is simple. Ducks do not have blood vessels or nerves in their feet so they cannot feel how cold it is.

Ducks can be observed eating gravel or sand. They store it in their gizzards and use it to break down their food. Ducks normally fly at low altitudes, between 200 and 4000 feet (61-121 metres) but an aeroplane struck a mallard flying at 21,000 feet(6.4km). It was bad news for the mallard, but it does posthumously hold the record of the highest recorded flight for a duck.

Geese are another domesticated birds that you can see on inland waterways, as well as on farmland. There are many different species. Some are migrants, such as pink-footed geese, which overwinter here from Greenland and Iceland, the barnacle geese which fly from Greenland and the North of Norway to Northern Britain during Winter, and greylags who fly here from Iceland. Two separate species of brent geese overwinter here- dark-bellied brent geese make the 3750-mile (6035km) journey from Northern Russia every Winter and settle in the East of Britain, whereas the pale-bellied brent geese fly from Greenland and Canada to Ireland every Winter. Canada geese were introduced to United Kingdom in the 17th century when James II imported some for his wildfowl collection in St. James's Park in London. As Canada geese tend to be very loyal not only to their partners but also to the site where they hatched, not many Canada geese were seen outside of the capital until the 1950s when the Wildfowl Trust distributed them around Britain. As they do not have many predators there numbers have grown, and they are especially suited to urban areas.

Geese use natural landmarks, celestial cues, and an internal compass to navigate these huge distances. They know when to migrate because the change in daylight length triggers a release of hormones which encourages them to gather. Before they set off, they eat a lot. Their body quickly converts food into fat, which is transformed into energy during their journey. The geese, like other migratory birds, wait for a period of calm weather before setting off on their journey. It is thought that this might be affected by climate change as more extreme weather becomes more common. Geese fly in a V-shaped formation and take it in turns to be at the front. Each bird is slightly above the one in front of them, which reduces air resistance. Scientists believe that geese honk in flight in order to maintain the integrity of the flock, and to co-ordinate the position shifts within the V formation.

Geese have much better eyesight than humans and can also see in ultraviolet light. They have a good sense of smell and can recognise different plants by their smell. They can also recognise individual geese, and goslings have been shown to recognise their siblings.

Swans are another bird which makes a long annual journey to the United Kingdom. It is possible to see three species of swan in Britain – mute, whooper, and Bewick's. Whooper and Bewick's swans are Winter migrants. The whooper swans fly to the U.K. from Iceland, travelling 600-800 miles (966-1287km) across the sea in a single flight. They fly at high altitudes and have been recorded flying at 8000 feet (2.4km), where the temperature was minus 50 C. They have a straight neck and a yellow beak, unlike mute swans who have an orange beak. Their names are onomatopoeic. Several years ago, I was walking my dogs at night when I heard their familiar whooping noise. I looked up to see a flock of whooper swans fly overhead, illuminated in the moonlight. Bewick's swans are one of the few birds in the United Kingdom named after a person, in their case the 18th century Northumbrian nature writer and illustrator, Thomas Bewick. These are the smallest swan that can be seen in Britain. The famous naturalist, and son of Scott of the Antarctic, Peter Scott studied Bewick's swans and realised that their beaks are as unique as a fingerprint. Bewick's swans migrate 4000 miles (6437 km) from Arctic Russia every Winter, although climate change means less of them are bothering to make the long journey. Mute swans, named because they are incredibly quiet during flight, are one of the heaviest birds in the world but can still fly at 50 mph. A seventh of the world's population of mute swans live in Britain. Mute swan feathers were used as quills. The Latin word for feather is *penne*, hence penknives which were originally used to cut quills.

In the Middle Ages every mute swan was the property of the local landowner. Annually the swans would be caught, and their beaks would be marked to indicate ownership. There were 900 such marks. Any swan without a mark was deemed the property of the Crown.

Until the 17th century swans were seen as a delicacy, eaten by aristocrats and monarchs, especially at Christmas. Swans have always been linked to death and transformation, as well as with the sun because of their association with the Greek God, Apollo.

You might have heard of someone's swan song, meaning their final performance before they retire. This phrase was first used by Thomas Carlyle in 1830 although the idea behind it comes from Greek mythology which alleged that mute swans sang as they were dying. Sadly, swans do not really sing.

Marsh harriers and ospreys are raptors who live in wetland habitats. Ospreys are a good example of a success story.[58] These raptors, which migrate annually to the United Kingdom, from West Africa, fly over the Sahara and the Bay of Biscay to reach Britain. It takes them about twenty

---

58 Although they are still rarer than Golden eagles

days, flying 261 miles (420 km) per day. Young ospreys fly to Africa in a south westerly direction and then stay there for three years, before returning to where they were born to mate. Naturalists still do not understand how ospreys are able to navigate so accurately but it is thought that, like pigeons, they use visual clues and geomagnetic field as well as the stars. As a result of persecution and the Victorian's obsession for collecting wild bird's eggs, by the First World War they were extinct in the U. K[59], but then in 1954 Scandinavian ospreys naturally recolonised Scotland, and in 2001 the RSPB released Scottish ospreys at Bassenthwaite Lake in Cumbria and Rutland Water in the Midlands. They can also be seen in North- West Wales. There are now about 250 pairs in the U.K.

The other name for an osprey is fish hawk and this is a good name for them because the majority of their diet is fish, so they live near freshwater or the coast. They have a coarse, scaled, and reversible hind toe which enables them hold onto slippery fish. I have watched them catch a fish at Bassenthwaite Lake in the Lake District and it is a magnificent sight. Although much slower than a peregrine they are still fast in a dive, reaching speeds of 78 miles per hour. They have a third eyelid which protects their eyes underwater. They do not dive down very deep, preferring to snatch a fish from just underneath the water surface. They normally take the fish back to their large untidy nests to eat. They can catch fish up to 2 kg (4.4 lbs). With larger fish they carry it with the fish's head facing the direction of the wind, in order to reduce wind resistance.

The marsh harrier was also persecuted to extinction, but like the ospreys, Continental marsh harriers naturally recolonised East Anglia and are now also in Southern and Western England, and Scotland. There are about 400 pairs. [60] They are not closely related to the other harrier species in the United Kingdom although they are the largest harrier. They have a long tail and quite distinctive V- shaped wings. Like the osprey they are migrants, arriving from Africa in April and returning in October. Marsh harriers have a quirky habit of changing their plumage so that some adults keep their juvenile plumage, and some older females develop a male plumage. A male mash harrier has more than one female which is

---

59 By 1840 in England and 1916 in Scotland
60 From 14 pairs in 1971

not an advantage because they often struggle to support all the chicks. I have seen marsh harriers are Blacktoft Sands in East Yorkshire, flying over the reed beds looking for small mammals and birds or frogs to eat. They fly higher than other British harriers so rely on their eyesight but also have fantastic hearing. Their face is like a giant ear, channelling sounds to the large ear openings behind their eyes.

Other birds that live in the reed beds are reed warblers and reed buntings as well as bitterns. Reed warblers only weigh 10-15 g (0.3-0.5 oz), about the same as a large envelope, but migrate from Sub-Saharan Africa to East Anglia and the South coast, with a few heading to Scotland and Ireland. They will make this journey two or three times during their lifetime. They have a distinctive churring call. They weave their nests between two or three reed stems.

Reed warblers are one of the species of birds duped by cuckoos into rearing their chicks, who are four times bigger than the unsuspecting surrogate parents. Reed warblers occasionally eject cuckoo eggs but only if the information is reinforced from a wider "neighbourhood watch." The cuckoo can make their eggs look identical to reed warbler eggs which means the reed warbler only discovers they are raising someone else's child when they suddenly have a huge chick in their nest.

Reed buntings are small sparrow-sized birds with a long-notched tail and white drooping moustache. They also like living on farmland, especially oil seed rape, but get killed by pesticides.

Bitterns are a brown wading bird, often seen shyly on the edge of reedbeds where they prey on fish, insects, and amphibians. These birds became extinct in Britain by 1870 before recolonising Norfolk before the First World War. However, in 1997 there were only eleven males in the United Kingdom. These are another success story. Following the RSPB's restoration of reedbeds bittern numbers have increased, and by 2019 there were 198 resident males, with another 600 bitterns overwintering here although they cannot be seen in Northern Ireland or the majority of Scotland.

Several years ago, I was at Blacktoft Sands in East Yorkshire. As my sister and I watched the bitterns in the reedbeds we heard a loud noise and thought it was the male bitterns booming to attract a female. Ten minutes later there was a tremendous storm. It was thunder we had heard, not the bitterns. This might sound as if we were being stupid but actually bitterns have the loudest call in Britain, and it can be heard two or three miles away and does sound a bit like thunder. Every bittern has a unique call.

There have been other ornithological success stories recently. Avocets are black and white waders which have a bill which curves backwards. They sweep their bills from side to side through the water, eating aquatic insects, worms, and crustaceans. You might think you do not know what they look like but if you have ever seen the R.S.P.B.'s emblem you will recognise an avocet because the R.S.P. B. used it as their emblem because it is one of their most successful conservation projects. In 1947 coastal marshes in East Anglia were flooded to defend the United Kingdom from potential invasion. This inadvertently created the perfect habitat for avocets who had returned to Britain after being extinct here for about a century. Now there are 1500 breeding pairs and 7500 winter migrants who live along the South and East coasts of England with half of them breeding at R.S.P.B. sites.

An even more recent colonist which, like the avocet, is equally happy in wetland or coastal areas, is the little egret. Its name comes from the French word *aigrette* which means *silver heron*. Egrets are small herons and have long feathers which cascade down their back during the breeding season. Between 1890-1921 there was a terrible Plume Boom during which it became very fashionable for ladies to wear large hats decorated with lots of feathers. Plume hunters were paid £2m per year to go round the world killing birds and selling their feather and pelts to milliners, mainly in London and New York. This led to the extinction, and near extinction, of many species of birds, including egrets. At the height of feather fashion, during the first decade of the 20th century, 14 million feathers were imported into the United Kingdom at a cost of £19m. By 1889 Emily Williamson in Didsbury near Manchester had had enough and set up the Plumage League. The largely female membership paid a tuppence annual membership fee and had to promise to discourage the wanton destruction of birds and not to wear feathers, except, for some reason, ostrich feathers. The group gradually gathered momentum and was given a Royal charter in 1904, becoming the Royal Society for the Protection of Birds (R.S.P.B.) It took many more years campaigning but, finally, in 1920 the society successfully persuaded the parliament to pass the Plumage Act which prohibited the hire, exchange, and sale of plumage and skins from certain wild birds, including egrets.

Another species of birds which benefited from this Act was the grebe family. These wetland birds, the most common of which are little grebe and long-crested grebe, have very elaborate courtship displays. They begin by shaking their heads and dipping their bills, before preening. Then they rush across the water to one another, chucking weeds at one another. If the weed is accepted, then the pair mate before building a fabulous floating nest.

Another bird which has made a recent comeback in Britain, after being extinct here for 400 years, are the United Kingdom's tallest bird, the crane. At 4 ft (1.2m) high its about 20% bigger than a heron. Like the bittern they have a loud booming call which can be heard two or three miles away. Although they like to roost between shallow open water and reedbeds or wet grassland and cereal stubbles these birds do not actually eat fish, much preferring a varied diet including butterflies and moths, snails and spiders, and small mammals and frogs.

Frogs are not the only amphibians to live in British freshwater habitats. Toads and newts can also be seen. One of the problems people have is differentiating between frogs and toads. Although they are similar it is relatively easy to tell them apart. Frogs have smooth, moist skin and long stripey legs, whereas toads are quite warty and have golden eyes. Toads crawl, unlike frogs who hop. Frogs have distinctive black patches behind their eyes whereas toads have large lumps, called parotoid glands, between their eyes. Toads can tolerate much drier conditions than frogs, and frogs need to spend a lot more time in water. Toads are normally just in the water in Springtime. There are two types of toad in the U.K- common and natterjack. Common toads are much rarer than frogs because of loss of habitat, pesticides, and getting squashed on the nation's roads. Natterjacks live in sandy habitats and are much rarer than the common toad. They can be differentiated by a yellow streak down the middle of their backs and their very loud croak, which the males make in the mating season in late spring and early summer. Their name literally means poison toad because natterjacks and common toads both have warts which are slightly poisonous, in order to deter predators; however, is also thought to refer to the noise they make, which can be heard over a large area.

If you see a clump of spawn it will be frogspawn, whereas toad spawn is in a line. If it is in double or triple rows it will be a common toad; natterjacks have a single strand. Frogspawn is actually 99.7% water. After anything between a couple of weeks or a few months, depending on the temperatures, the frog or toad spawn will hatch into tadpoles. Tadpoles have fish hearts, whereas frogs have an amphibian heart. There can be up to 2000 tadpoles in a clump although only an exceedingly small number of these will become frogs or toads because tadpoles are an important food source. In the Middles Ages tadpoles were believed to cure gout, but I would strongly advise not to eat them.

Frogs can lighten or darken their skin to fit in with their surroundings. They do not drink but, instead, absorb water through their skin. They can also breathe through their skin as well as their lungs. Frogs have long, sticky tongues which they use to catch worms, slugs, snails, and flies. As frogs are cold- blooded they need to hibernate over Winter, normally under logs or at the muddy bottom of a pond. As the temperatures start to rise in early Spring you should start to see them again.

In the Middle Ages frogs were worn around the neck to stop the Devil from entering the body, in the hope he would choose the frog instead. Toads were often linked to the Devil. At York Minster there is a 11th century Doomstone depicting poor sinners in a pot over a fire which is being stoked by demons. There are numerous toads on the stone, coming out of people's mouths and lurking near the demons.

There are three species of newts in Britain- smooth, palmate, and great-crested. Smooth newts are nocturnal and are more likely to be seen under stones and in compost heaps. They hibernate between mid-October and February or March, emerging once the temperature gets over five degrees Celsius. During the breeding season smooth newts become more colourful, and preform an elaborate courtship display – crests erect, tail curled, wafting his pheromones towards the female. If she likes the smell of him, she will mate with him. Palmate newts get their unusual name from the male's hind feet during the breeding season. Palmate means the webbed bits between the palmate newt's toes. Great-crested newts are the rarest of the British newt species. Their black spotted bellies are as unique as fingerprints. They do a strange dance to attract females. They can excrete a foul-tasting substance from their glands to ward off predators. They also have the ability to sniff out fish-free ponds, and other sources of water, and can determine if the water has a neutral pH, which is preferable to lay their eggs in. Female great-crested newts lay 200-300 eggs, two or three at a time, in the freshwater. She individually wraps every tiny egg in leaves to protect them from ultraviolet light and predation. After four weeks they become tadpoles, and then four months later they turn into "efts". The efts are then able to leave the water, living off slugs, worms, and insects. After up to three years on land they will be sexually mature. They can live up to 25 years, but unfortunately their numbers have declined dramatically because of pollution and habitat loss, although they are a legally protected species.

Many insects rely on freshwater, providing food for many of the other wetland dwellers. In early Summer I like to dip my hot, tired feet in a Cumbrian stream and watch dragonflies and damselflies dart up and down the river. The species are similar, but it is relatively easy to tell them apart. Dragonflies are bigger. Damselflies have four uniform wings, whereas dragonflies have slightly smaller wings at the front. At rest dragonflies keep their wings open, whereas damselflies hold their wings back along the length of their body.

Dragonflies have the largest eyes of any insect in the world. Eighty per cent of their brains are used to process what they see. Humans have tri- chromatic vision (which means we can see a combination of red, blue, and green). Most of the animal kingdom sees in di-, tri-, or tetra vision. However, dragonflies have between eleven and thirty opsins, which are light sensitive proteins in their eyes, which means they can see colours that we could not even imagine. They have a median line, like a cross-hair, to enable them to lock onto prey, and they can process information from each eye individually at a speed of 30mm per second. In addition, they have sixteen neurons in their brain which control their flight motor in their thorax. This enables them to work out the trajectory of their prey so that they can be caught on the wing from behind. They catch prey feet first, ripping of the insect's wings before eating it, all in flight. Their favourite prey are midges, and they can eat hundreds per day, which is good for me because midges like to bite me. In contrast, damselflies have binocular vision which enables them to fuse images together and to judge distance. They catch prey from the front.

Damselflies and dragonflies can control their wings independently so they can fly up or down, sideways, and even backwards. They can also hover for a minute and reach speeds of 30mph.

Dragonflies lay their eggs under water where they exist as nymphs for two years. Nymphs look like adult dragonflies but do not have wings. They have hinged jaws to enable them to catch prey. Unlike the adult dragonflies, the nymphs are quite dowdy because they need to be camouflaged, to try and avoid predators.

Despite their names mayflies can be seen at any time of the year. Out of the 52 species in the United Kingdom only one can be seen in May. Mayflies begin life as a worm, living in mud for three years before emerging one sunny day, mating, and then dying just a few hours later. Uniquely they have two adult stages. They emerge from the water as nymphs, then sheltering on the riverbank, where they molt again, replacing their drab skin with a new shiny one.

Caddisflies are another species which you are likely to see at dusk or at night. They are moth-like insects with hairy wings. There are about 200 species in the United Kingdom, making them one of the most diverse freshwater species in Britain. About 50% of them are seen near ponds, with the most common being the thirty species of cinnamon sedges which you might see over your garden pond in the evening.

They start life as larva underwater. They make a case by gluing together with silk bits of plant, grains of sand, and even snail shells. This helps to protect them from predators. Caddisflies are popular sources of food for many freshwater species including Atlantic salmon, brown trout, and dippers.

Of course, one of the things that you are likely to see in freshwater, whether it is a river, lake, or pond, are fish. I suspect that unless someone is a keen angler, they probably do not think about fish very often, I certainly do not, but they are actually fascinating. Over 530 million years ago fish were the first animals to evolve bony skeletons. There are more than 34,000 species of fish in the world, more than all the vertebrates in the world combined. New species are still regularly being found.

Fish have good eyesight, being able to differentiate between colours and see things as small as plankton, but only in clear water and in daylight. In deep water, and at night, they only see in black and white. Their eyes are on the top of their head so they can only see prey above them. Although they have peripheral vision, they cannot see directly behind them.

Fish also have a sense of smell which they use to find food and differentiate whether the thing in their mouth is food or whether it is something they should spit out. Scientists now think that fish, such as trout and salmon, use a chemical scent trail to help them return home to spawn, which is when they mate and the female lays eggs. She will have up to 10,000 eggs, but these are such a tasty snack for so many other water dwellers that only one egg is likely to become another fish.

The 28,000 species of bony fish hear by using bones in their heads called otoliths. There are three pairs of otoliths- the sagittae, the lapilli and the asteriscii. They are made from calcium carbonate, like a shell, and differ in size and shape depending on the species of fish. These grow throughout the fish's lifetime (but are still only the size of a grain of sand in juvenile fish and the size of a fingernail in adult fish) and have rings, like the ones you would see on a tree stump. These rings can provide scientists with a huge amount of information because the otoliths contain chemical elements from the water they have swum in. The otoliths are like the fish's 'black box' (data recorders on aeroplanes) but also allow the fish to keep their balance and to detect water depth. By examining the otoliths of fish scientists can work out where the fish have been, how long they have spent in various places, and how fast they have grown. This helps them collate information on fish migration. Just like birds, some fish, like trout and salmon migrate huge distances to return to the rivers of their birth to mate and spawn. Salmon migrate 1864 miles (3000 km) Some species of fish migrate from oceans to freshwater, and some swim in from freshwater to seawater. It has often been wondered by scientists how fish migrate. As well as using chemical scents it is thought that species like salmon, similarly to birds, use the earth's magnetic field, but research into this is ongoing.

Fish have nerve endings on either side of them, called the lateral line, which help them detect motion and determine an object's size and speed and the direction it is moving in. The lateral line acts like an in-built underwater radar system.

You might have held a fish and noticed that it feels a bit slimy. This is because fish secrete a mucus which protects them from parasites and disease, helps them fight infection, and ensures they move faster through the water. Some also release toxins to deter predators.

Many species of fish have taste buds all over their body. Fish communicate by making a series of low-pitched sounds with various parts of their body. They can use chemicals in their skin to change colour, so they are camouflaged according to their habitat.

Fish breathe through their gills, which are thin sheets of tissue richly supplied with blood vessels. As the water passes over them oxygen is absorbed into the fish's blood, and waste products, such as carbon dioxide are released into the water.

Two things which might surprise you about fish are that fish can drown if there is insufficient oxygen in the water, and about a third of British fish actually change gender because of human sewage pollution discharged into our river system.

There are numerous native British fish, but the following are some of the fish you might see in British rivers and lakes. Pike are freshwater predators. They get their name from the weapon which was commonly used during the British Civil War. These extremely versatile fish can live in a variety of freshwater habitats. They lie in wait for their prey, mainly fish and frogs, then pounce on the unsuspecting victim.

Grayling and brown trout share habitats because they both eat slightly different prey. Their presence in a river is a good indicator of clear, clean freshwater. Brown trout are the most genetically diverse vertebrates in the United Kingdom with at least fifty species. To complicate it further brown trout has been known to interbreed with salmon so there are various hybrid species too. They have 38-42 chromosomes, far more than humans who have 23 pairs. Their elliptical eyes mean that their eyes act independently so they can focus on food and keep an eye out for predators at the same time. They need temperatures under 20°C. They change colour to fit in with their habitats and to show how aggressive they are, with darker trout being the most aggressive. Not all trout migrate to the sea, some just migrate up and down rivers. Some trout set off and then change their minds. Scientists do not know why this happens and research into it is ongoing.

Atlantic salmon are the most common species in British rivers although alarmingly their numbers have halved in a decade due to pollution, habitat loss, and over-fishing. The presence of salmon is an indicator of an exceptionally clean river. They are large fish. A two-year-old salmon is normally about 30 inches (75cm) long. They are blue-silver with black dots on their back and some have slightly pink bellies. They eat insects, invertebrates, and sometimes plankton. Adult salmon mainly eat small fish, especially capelin. They hatch in freshwater and after several years migrate to the sea, before eventually returning home to mate and spawn in their home river. They die soon after breeding. It is known that they can leap over large obstacles to get back home, jumping up weirs, dams, and waterfalls. One was recorded jumping ten feet (3m).

Chub like deeper, slower moving streams and use their big teeth to break down fish. Tench also like soft -bottomed lakes and slow-moving rivers. They used to be called doctor fish because they were mistakenly thought to have healing properties. Medieval doctors used to mulch them up and cover people with their slime. I would not recommend it.

When I was a child, I used to love dipping a net into streams. The two fish I usually caught were minnows and sticklebacks. Minnows are a favourite food for many birds and fish. Luckily, they are abundant in well-oxygenated streams and rivers. They are tiny – about the size of a little finger- and change colour in the breeding season in spring, with the males developing a red belly. Sticklebacks get their name from the Old English word *sticei* which means spike. They have between three and nine spines. The male creates a shallow 'nest' in the pond or stream's bed and fills it with small plants, which they glue together with a fluid which they excrete from their kidneys. He constructs a tunnel, which is marginally bigger than him. If he successfully finds a female stickleback, he persuades her to lay her eggs in the 'nest'. He then cares for them by fanning them with his tail to ensure a flow of freshwater over them. When the eggs hatch, he will care for the 60-80 fry by himself.

I also remember seeing a rarer British fish in my local river when I was a child. Although it was forty years ago now, I still remember the mixture of fear and excitement when I saw what I initially thought was a snake. Eels do look snake- like, but they are an elongated fish growing up to 13 ft (4m) long. They swim by generating body waves which travel the length of their bodies. If they reverse the direction of the wave, they can swim backwards, the only British freshwater fish which can do that.

What I did not realise forty years ago as I watched the eel wriggle around in the water near the ford, was that that eel, and every other eel, comes from the Sargasso Sea, part of the Atlantic Ocean, and that it had gone through many transformations before arriving at the river Leven. They start life as flat, transparent larvae called leptrocephall (a good word in scrabble) which follow the Gulf Stream for at least a year before reaching European coastal areas where they turn into glass eels (juvenile eels). Glass eels are so called because they are transparent, although they have red gills, and their heart is visible. On entering the British river system, they transform again into elvers. At this stage their pigmentation darkens, and their skin thickens. The elvers must determine their gender at this stage. Only female eels will return to the body of water where their ancestors lived, using their incredible sense of smell to trace the DNA left by their elders. It is safer for the males, who just return to live in the sea, but the advantage of being female is that they get to pass on their genetics to the next generation. On reaching their destination they become adult eels. Adult eels live at the bottom of rivers, canals, ponds, and lakes, eating worms, insects, fish, and frogs at night, as well as scavenging on dead animals. Unlike salmon they like a body of water with plenty of silt. Between the ages of eight and eighteen they will return to the Sargasso Sea to spawn and then to die. Before the journey their appearance changes again. Their eyes grow larger because they spend part of the journey deep under the water. They fatten up and become a silvery-white. Their anus closes, maybe to prevent infection, so they cannot eat on the 3107-mile (5000km) journey. They do not use a direct route back, instead using the subtropical gyre system (a circulating ocean current) that flows to the Caribbean.

European populations of eels have reduced by 80% partly due to a shift in the Gulf Stream, man-made barriers, and a parasite which affects the eels' buoyancy aid. However, in recent years their populations have started to recover, partly due to the warmer temperatures caused by the North Atlantic Oscillation Index, sending the leptrocephall in the right direction again. Hopefully their populations will continue to increase because eels have been around for seventy million years.

One of the strangest fish you might see in our rivers is the brook lamprey which is a primitive jawless fish. At their larval stage they use suckers on their mouths to eat bacteria and algae off the bottom of waterways, so are elusive. At that stage they do not have eyes. They live in the larval stage for three to seven years before burying in the sand and metamorphosing into adults. When they transform into adult they develop eyes, but they stop feeding. The adults migrate upstream in the spring, before spawning then dying a few weeks later.

Wetland areas are also important habitats for many species of plants. One of these which I love because of its pretty pink- red flowers is ragged Robin which loves damp, boggy, and marshy conditions. Before Linnaeus came up with his binomial system, like bird names, different counties had different names for wildflowers. Many counties seem to have used the word ragged, because of the appearance of the flowers, with the plant being called ragged Jack in Essex and ragged urchin in Devon. In Yorkshire it was called thunderflower because it was thought that if you picked it and brought it into the house there would be a storm. It might be called ragged Robin because of the colour of its flowers, but it is thought that Robin, or Robert, in a flower's name refers to Robin Goodfellow, who you might know as Puck. He was a mischievous goblin popular in Medieval English folklore. He appears in literature including *A Midsummer Night's Dream*. The genus Latin name is the Greek word *lychos* which means lamp because ragged Robin was used as a wick for oil lamps. Ragged Robin is a fantastic pollinator, and its flowers were used as shampoo, and its roots can be made into a soap substitute for washing clothes.

Marsh Marigolds are a big buttercup- flower which like marshy ground. It is also known as Mayflower (or May blobs where I am from) or Kingcups. It supposedly gets the latter name because the flowers look like the cup from which Kings drunk from. This is reflected in its Latin name *Caltha* which is derived from the Greek word for goblet. This plant is popular with pollinators and grows from March to July. It is thought to have been here since the Ice Age. It has several medicinal uses including being used to treat bronchial and menstrual problems, but be careful it is poisonous so do not pick it yourself. [61]

Another pretty yellow flower, one of the first spring flowers to come through, is coltsfoot which has been used for at least two millennia to treat respiratory problems, and ulcers and inflammation in livestock and humans. It also helps to improve the immune system. It gets its name from the hoof-like shape of its leaves.

Bistort is a red flower which can be seen in damp habitats. It gets its name from the Latin *bistorta* which means *twice twisted* and refers to its twisted root, which was used in the Middle Ages to treat snake bites. It was used to treat mouth and throat infections, diarrhoea, colitis and cystitis, and as a poultice for wounds because it helps staunch bleeding. In Cumbria it was used as part of a traditional Easter Pudding, along with nettle tops, dandelion leaves, and Lady's mantle, which was eaten with meat on Easter Sunday. It was mixed with frankincense to heighten the psychic powers of clairvoyants.

Two of my favourite flowers are lady's smock and bogbean. The former is a common wildflower, often seen in damp meadows, and is also known as cuckoo flower because it can be seen in April, May, and June. Unlike other flowers with Lady in the title it is not named after the Virgin Mary but gets its name because the mauve coloured flowers are the same colour as upper-class lady's dresses in the Middle Ages. Sadly, bogbean is much rarer but it can be seen in some areas, such as upland bogs, fens, and lake margins, in May and June. It is an incredibly beautiful delicate white flower. In the Middle Ages it was used to treat flu and aid digestion but do not pick it, it is a protected species.

Many mint species like wet conditions including peppermint and water mint which were both used to treat stomach, liver, gallstone complaints as well as being used to ease toothache and prevent nausea. Both have purple flowers and have leaves which smell, unsurprisingly, of mint. Amongst this family is also gypsywort which likes reedbeds, ditches and

---

61 If you would like to use an herbal remedy always seek expert advice first.

overgrown pond margins. It has a long stalk with interspersed clusters of white flowers. It gets its name because it can be made into a dark dye which was supposedly used by gypsies to tan their skin.

Butterbur grows on riverbanks and damp waste ground between March and May. You might have noticed their large leaves, which were once used to stop butter from getting rancid, hence the name. This plant was also used to treat coughs, urinary problems, lower back pain, intestinal problems and as a painkiller and to reduce bleeding.

In Spring and Summer look out for various species of marsh orchids in marshy areas. You might also be lucky enough to see a rare plant which looks too exotic to be growing in the British landscape, marsh helleborine. This pretty flower is a good indicator of species-rich fens and grasslands.

A less welcome plant alongside British rivers is Himalayan balsam which was introduced into the United Kingdom in Victorian times as a garden plant before escaping and growing wild throughout the country. Although this plant has sweet smelling pretty pink flowers which are popular with pollinators, it is treated as an invasive species because it has pods containing many seeds. My sister and I used to like popping them when we were children. I now wish we had not done so because these plants take over riverbanks, sucking up nutrients, stopping other plants from thriving. The seeds can travel up to thirteen feet (four metres). The plant grows quickly, soon reaching eight feet (2.5m), blocking out light for smaller plants. When it dies back in the winter it leaves riverbanks open to erosion and its dead leaves block waterways and cause flooding. It is an exceedingly difficult plant to eradicate because it can stay underground for a long time and the slightest amount of light can regenerate it.

On top of the water, you might see river - water crowfoot which is another member of the buttercup family and grows from May to August. You might not know what it looks like but if you have ever seen John Everett Milas's painting of Ophelia in her watery grave you might have noticed it. It is the small white flower spread around her. Both trout and mayflies like this flower. Unfortunately, it is now rare, although more common in the Welsh Borders and the Midlands. The chalk-stream loving, stream - water crowfoot, is even rarer, only growing in some chalky shallow streams in the South of England.

The Wonders of the Wild Places

More common are water lilies which like slow or still nutrient- rich water with muddy bottoms. The flower can grow up to ten feet (three metres) in diameter. In her book *Hidden Nature* Alys Fowler likens waterlilies to being like a forest because their roots draw oxygen down to the bottom of the waterway, thus allowing other species to thrive, and their leaves provide shade for fish, shrimps, and other water dwelling species. Their leaves contain spongy cells which trap air. This acts as an internal life jacket, keeping the plant buoyant.

Yellow irises, also known as flag irises, because there outer petals droop like a flag at half mast, are often a colourful addition to wetlands, growing on the margins of waterways, ponds, fens, and reedbeds between May and August. However, they sometimes take over a small pond so be careful about planting them in a garden pond. The name *Iris* is the Greek word for rainbow and these plants come in a variety of colour although in this country you are most likely to see them in yellow, purple-blue, and white. Iris, the Greek God of Rainbows was the messenger between heaven and earth. Purple irises were planted over the graves of women to summon Iris to guide them to heaven. Irises have many meanings depending on the colour of the iris. Purple irises are linked with royalty, whereas the flag iris is symbolic of passion. Blue irises are symbols of faith and hope.

In wetland areas you will also often see reeds, which are an important habitat for rare species such as bitterns and marsh harriers. The most seen reeds are unsurprisingly the common reed which, like bracken, has rhizomes which spread a long way, creating dense stands of reeds. In places like East Anglia, the reeds were harvested for thatch, which helped to maintain the reedbeds. However, reedbeds began to be converted into agricultural land in the late 17th century. A decline in reed cutting as well as pollution and agricultural run-offs have led to many reed beds being in an extremely poor state. So much so that of the 900 reedbeds currently in the United Kingdom only fifty are big enough at the present time to support species such as bitterns, which need twenty hectares of reedbed. However, things are beginning to improve, and nature organisations and landowners are beginning to release the importance of properly maintained reedbeds. The Wildlife Trusts alone have planted 1500 hectares of new reedbeds in recent years. This will have a huge impact on wetland species, although there are still concerns about the effect of climate change, especially on coastal reedbeds.

Some species of trees prefer wet condition too, especially alders and willow. Trees near a river or wetland habitat provide shelter for fish, mammals, and birds; provide shade, reduce water temperature, maintain oxygen levels; act as a buffer filtering pollution and run-offs; and their roots stabilise riverbanks and reduce erosion.

Alder's wood does not rot when it gets waterlogged; instead, it actually becomes stronger. In doing so the roots help stabilise riverbanks and reduce flooding. It is the only British deciduous tree to have cones which are actually the female catkins which stay on the tree all the year round. In spring you will also see the yellow male catkins. Alder is an especially important species. Not only does it provide a home to a variety of other species, including siskins and goldfinches, craneflies, otters, and chequered skipper butterflies, but it also has a very valuable symbiotic relationship with a type of bacteria *Frankia alni* which lives in the tree's large nodules in their root systems. Alder, with the help of the bacteria, draws nitrogen from the atmosphere, and from bird and animal poo, and converts it into fertilizer, thereby improving the soil quality.

Like other plants alder has several practical uses. Its wood was used to build boats and make clogs. A piece of alder in a drawer was meant to deter woodworm. Alder leaves in someone's shoes were said to cool feet and reduce inflammation. The flowers produced a green dye- think Robin Hood's outfit- which was said to be favoured by fairies to colour their clothes.

When the trees are cut, the pale wood turns deep orange, making it appear to be bleeding. Therefore, in Ireland it was thought to be unlucky if you passed an alder on a journey.

Willow was introduced to the United Kingdom by the 18th century poet Alexander Pope, who was also a keen gardener. There are lots of species of willow, further complicated by many hybrids as well. There are sixty hybrid varieties just of the osier willow alone. This willow has a useful ability. It can absorb heavy metals and is used to decontaminate waste ground. White willow is the largest species and a popular variety of it, *Salix alba* 'Tristis', is better known as weeping willow. Crack willow gets its name from its habit of splitting and the noise its branches make when they snap off. Goat willow is better known as pussy willow because its male catkins look like cat's paws. It can live up to 300 years.

Willows have many uses. The best known are probably cricket bats and aspirin. The former comes from a hybrid of white and crack willows, also called the cricket bat willow. The wood is taken from the female trees. The latter is made from salicin which is found in willows and other plants too. Its botanical use is to promote growth and reduce stress. Since at least Ancient Greece, the bark was chewed to release the salicin because it was known to be pain relief, as well as used to treat inflammation caused by arthritis, and made into a liquor to treat diarrhoea, stop bleeding, and clean wounds. Incidentally salicin is Latin for willow. The Reverend Edward Stone was the first person, in 1763, to provide the first modern scientific description of the medical use of willow bark[62]. Medical advances allowed Henri Leroux to isolate salicin from the bark in 1826. In 1853 Charles Frederic Gerhardt created acetylsalicylic acid, made from modified salicylic acid from willow bark, for the first time, but neither used it nor marketed it. Meanwhile, physicians began prescribing salicin to treat people with rheumatic pain. Then in 1897 Felix Hoffman, a German chemist, also made acetylsalicylic acid, which he called aspirin. However, his colleague, Heinrich Dreser, was dismissive of this new drug, believing it enfeebled the heart, although the real reason might have been that Dreser was too busy creating heroin, which was to be marketed as a cough remedy. Eventually Dreser came round to the benefits of aspirin, but only after he had tested it on his rabbits and himself. It was not until the beginning of the 20th century that Lawrence Craven began to realise that this was a useful drug in the treatment of heart problems, especially to prevent myocardial infraction and prevent strokes. Unfortunately, the discovery did not help Dr Craven who died of myocardial infraction in 1957. However, it has probably saved the lives of millions of people around the world. Aspirin has recently also been found to reduce some cancerous growths so might play an important role in fighting cancer, so I suspect that this incredibly special drug, made from willow bark, might save many more millions in the future.

---

62 Greek physicians Galen and Hippocrates were the first to record its uses.

Crack willow's slender, flexible stems were used for basket weaving and the larger stems were made into boats. Pussy willow was made into wattle and used to build houses. Osier willow was used to make baskets and is used to make willow sculptures. Willow wood was also used to make prosthetic limbs.

Willows are an important habitat for many other species including many species of moths and butterflies, especially the rare purple Emperor butterfly, eyed hawk-moth, and willow ermine.

Some species of mammals also like living in freshwater habitats including the U.K.'s fastest declining species, water voles, which have declined by 94% since the millennium. When I was a child in the 1980s, I regularly saw these by the river as I left my ballet lessons. I used to be frightened of them because Kenneth Graham in his wonderful children's book *Wind in the Willows* had a character called Ratty, which is actually meant to be a water vole. Water voles have a rounder face than rats and do not have the long pink tail.

Loss of habitat has been one of the reasons for the sharp decline in this vole, but another riverside mammal is also to blame. Mink love eating water voles. This species is non-native to the United Kingdom and was imported into Britain in 1929, when fur coats were popular. Mink were bred in fur farms until escaping in the 1950s. Their numbers have increased rapidly in the last forty years and most of the rivers near me now have mink. They look a bit like an otter but are a deep brown and smaller and slimmer. They also eat fish, small birds and mammals, and invertebrates. They are voracious predators, often killing a lot more of other species than they need just for food. They can devastate populations of ground nesting birds as well as water voles.

Unlike mink, otters are native to Britain and have been here for millions of years but were driven to the brink of extinction in the 1970s because of persecution (anyone who has wept their way through Tarka the Otter will know what that involved) and dangerous pesticides. However, this is one of the species which has increased in numbers in recent numbers, particularly due to cleaning up rivers, and now they are widespread across Britain. They have large territories. These brown members of the mustelid family have paler necks and chins and very noticeable whiskers. Most of their diet is fish, but they will eat birds, mammals, and frogs too. They have an acute sense of smell, hearing and eyesight and are strong swimmers, although they can only hold their breathe under water for about thirty seconds so you can sometimes see rings on the water or see them on the riverbank or shoreline. Fortunately, otters might help reduce mink populations because they tend to live in the same habitats and eat mink kits and occasionally adult mink too. The best time to see otter are at dusk.

Another mammal has recently made a return to British waterways after being extinct for centuries. In 2009 Norwegian beavers were released in Knapdale, Argyll. In England beavers were discovered on the River Otter in Devon in 2015 and have also been released in Cornwall and Kent as well as Wales.

Beavers are known as a keystone species because they can create habitats for other species. They have already created 13,045 square miles (33,670 square kilometres) of new freshwater habitat, the equivalent of ten Olympic size swimming pools. In her fascinating book *Wilding* Isabella Tree asserts that beavers might play an important role in mitigating climate change because the ponds they create act as carbon stores.

They are Europe's largest native rodents. According to the Scottish Beaver Trust they are a similar size to, "a tubby spaniel." They are one of nature's best engineers, coppicing trees, and building dams which create ponds and wetlands, and allowing water to move more slowly, meaning it does not stagnate, leading to greater biodiversity. The Wildlife Trusts describe them as "ecosystem engineers," and this describes just how important they are as creators of new habitats. Their dams are leaky, creating changeable habitats which reduce flooding, increase water retention, improve water quality, and reduce siltation and pollution.

It is not only mammals which benefit from wetland habitats, they are also important habitats for lots of small species of insects, as well as the mis-named grass snakes and native white crayfish.

Grass snakes are non-venomous green-grey snakes, with a yellow and black collar and pale belly with dark markings on the side. They are the longest snake in the United Kingdom, getting up to nearly five feet (1.5m.) Look out for them, especially in garden ponds in the South of England. These snakes are often found in or around water because they feed on amphibians. They like laying their eggs in people's compost heaps, up to forty eggs at a time. Be careful if you come across a grass snake because if they feel threatened, they will play dead and emit a foul-smelling substance, which from what people have told me, is really unpleasant, and difficult to get rid of.

White-clawed crayfish are the only native crayfish species in the United Kingdom. About a quarter of the world's population live in Britain. They are an important food source for other wetland creatures. When they are young they are eaten by dragonfly larvae, when they are adults they are eaten by trout, herons, mink, and otters. Unfortunately, these crayfish are in decline because of a foreign invader, the American signal crayfish.

Some of the smaller species you might see if you dip a net into a freshwater habitat are pond-skaters, freshwater shrimps, and water boatmen. There are ten species of pond-skaters in our ponds, rivers, and lakes. They are also known as various names including water striders, water skeeters, and Jesus bugs. This innocuous looking creature, with its long spider-like legs, is actually a predator who detects the vibrations on the water's surface to detect small insects which it stabs with its sharp mouthparts. It feeds on these insects as well as scavenging on dead animals it finds in the water. They have water-repellent hairs on the bottom of their feet which allows them to walk on the surface of the water.

Freshwater shrimps are less than 0.8 inches (21 mm) long. They are a greyish or brownish freshwater crustaceans and are not really shrimps. They have a complex lifecycle. A male carries a female around for several weeks before mating with her, when she is finally released. After a month, young shrimp can live independently and after a few months they are able to breed. As female freshwater shrimp can produce fifty babies each month, the population grows very quickly. They prefer well-oxygenated water. They are an important food source for many freshwater creatures so spend a lot of time hiding near rocks and vegetation.

There are two types of water boatmen in the United Kingdom- lesser and greater- although the greater water boatman is also known as the common backswimmer[63] and the lesser water boatmen swim on their

---

63 There are actually 4 species of backswimmers.

front. Like pond-skaters they are seen on the surface of the water and are voracious predators, eating insects which have had the misfortune to fall into the water. They also eat tadpoles and even small fish, sensing the vibrations in the water before stabbing them with their 'beak' before injecting their victim with a toxic saliva, and sucking out the contents of the victim's body. For this reason, be careful if you have one in your net- if you touch it, it might give you a nasty bite.

Another animal which you might see in freshwater in the United Kingdom, although it has become exceedingly rare in recent years, is the medicinal leech. These are the largest species of leech in the United Kingdom, growing up to eight inches (20 cm), and the only one which feeds on human blood. It is dark brown-black with thin yellow, green, and red stripes down their sides. They have two suckers – a posterior sucker – which they use for leverage- and an anterior sucker, which contains three jaws which between them contain a hundred teeth. Despite liking human blood, they normally feed on the blood of amphibians, fish, birds, and mammals, with amphibian blood being their preferred choice. They are opportunistic feeders, able to exist for months without food. They like shallow ponds and ditches with high and fluctuating water levels.

Leeches have been used in medicine for millennia. Ancient Egyptians prescribed them for everything from fever to flatulence and the well-known Roman writer, Pliny the Elder recommended them as a treatment for piles. The Ancient Greek doctor, Hippocrates, believed that the body had four humours – blood, phlegm, yellow bile, and black bile- which were centred in specific organs – the brain, the lungs, the spleen, and the gallbladder- which were related to four specific personality types – sanguine, phlegmatic, melancholic, and choleric. He believed that an imbalance of these humours caused ill health and therefore advocated removing the excessive humour by various methods including blood-letting. Combined with the writings of Galen of Pergamum who became convinced that blood was the most important humour, this became the basis of medical treatment for centuries, irrespective of the condition. When Charles II had seizures and George Washington caught a fever they were both prescribed leeches. They both died.

Then in the 18th century a French physician called Francois-Joseph-

The Wonders of the Wild Places

Victor Broussais began treating Napoleonic soldiers with leeches. This started a 'leech-craze' across Europe and North America leading to a huge demand for leeches. Partly thanks to Dr. Broussais, France alone required 35 million leeches per year. Leech shapes appeared on ladies' dresses of this period and the well-to-do impressed their friends by keeping jars of leeches on their mantlepiece. Jars of leeches in rainwater were seen in every doctor's practice and chemist shop. Across Europe leech hunters waded through shallow ponds hoping to attract leeches. Such was the demand for leeches, medicinal leeches became almost extinct in many European countries, including here, and were probably only saved by people using a leech and then throwing it into a ditch. Leech farms were set up, with cattle and donkeys being used to harvest leeches. However, this was a risky business, because these large overcrowded, poorly oxygenated pools of leeches increased disease and attracted predators. One leech-farmer lost all his 20,000 leeches in one day when they were eaten by a hungry flock of migratory ducks.

Leeches were not without their problems, despite their popularity. It was difficult to judge how much blood they consumed (although it is now known it is between 5-10 ml of blood – ten times their bodyweight) and the person continued to bleed once the leeches were removed so a lot of people died from blood loss. Also, leeches were reused thereby leading to infection. For this reason, an inventive French physician called Jean-Baptiste Sarlandiere created a mechanical leech which replicated the actions of the leech and were popular with doctors because they did not get ill or die, could suck more blood, and were easier to transport. However, despite this, real leeches, by this time often imported from Russia or Hungary, continued to be used.

The 'leech-craze' began to wane in the 1830s when doctors realised that leeches were ineffective for treating cholera which was sweeping across Europe. By the end of the 19th century many European doctors were no longer using leeches, but in the U.S.A. they were used regularly well into the 20th century.

However, that is not the end of the story for medicinal leeches. Whilst doctors like Hippocrates and Galen were wrong about humours, leeches do have many medical uses. When they bite they inject an anaesthetic into the skin. Their saliva contains hirudin, a natural anticoagulant. In the 1970s leeches began to be used by doctors again. Surgeons realised that leeches were useful in reconstructive surgery and microsurgery. For example, if a finger is severed it is difficult to repair the smaller veins, so the finger becomes swollen with blood. Leeches help by sucking out the excess blood steadily, allowing time for the smaller veins to knit together and heal, and the anticoagulants the leeches inject into the skin prevent blood clots. In 1985 an American physician called Joseph Upton also successfully used leeches to help him re-attach a five-year old's ear which had been bitten off by a dog. Beforehand this would have been impossible because of the tiny capillary connections, thanks to leeches these sorts of operations are now normally a success.

Thankfully, the leeches used in modern medicine are not taken from the wild and the poor do not need to walk around in shallow ponds trying to catch them. All leeches used in medicine today come from a firm in Swansea who harvest leeches on blood-soaked sausages before exporting them all over the world to be used in surgery. Sixty thousand leeches are exported from Wales every year.

Snails are another species you might see in and around water which could have an incredible effect on modern medicine. Snails and slugs secrete visco-elastic slime which enables them to adhere to, and glide over, surfaces. It also prevents them from drying out as well as making them less attractive to predators. This slime is extremely complex and goes against Newtonian principles because it acts like a solid glue at rest but liquefies on force, very much like non-drip paint. However, once the force is removed the slime re-solidifies. This process is known as adhesive locomotion, and snails use it to keep part of it attached to the surface whilst it moves forwards.

Incredibly, scientists at Heriot-Watt University in Edinburgh believe that this slime could be used to heal broken bones and could be extremely useful, especially in hip operations. David Nunn, a Consultant Orthopaedic surgeon at Guy's Hospital in London believes that this could be especially useful because it could help accelerate bone healing when natural healing has not been achieved. This special mucus, in this case from the dusky arion slug, has inspired 'bio-glue' which is extraordinarily strong, moves with the body, and sticks to wet surfaces. It also has a shock absorbent component which is crucial because it takes the strain, so the adhesive stays stuck. The glue is non-toxic and is three times stronger than existing medical adhesive. A team at Harvard University have successfully used the glue to seal a hole in a pig's heart so it might eventually be able to be used in human heart surgery. It is also being looked at whether it could be used as a patch or an injected liquid to release drugs into the body, and to keep things like pacemakers in place. The Wyss Institute in Boston, U.S.A, believe that it would be cheap to produce. They had applied for a patent, but I have not been able to find out whether it is currently being used in surgery in America or Britain.

As if bio-glue is not enough, remarkably snail and slug slime also have other medical and cosmetic uses. Hippocrates first advocated snail slime for healing wounds and ulcers and reducing scarring. Then in the 1980s Chilean snail farmers realised that when they cut their hands, the wounds healed quickly, and that their hands felt soft. This led to a $314m global skin products industry, which are particularly popular in the U.S.A. and South Korea. It is claimed that snail and slug slime make skin softer, can be used to treat acne, and reduces scarring and wrinkles. This is not just the latest fad. There is some science behind it, which might help treat people with skin problems, burns, and warts as well. Although slug and snail slime differ slightly, depending on the species, it includes glycoprotein enzymes and antimicrobial peptides[64], such as mucin which has antibacterial properties. These stimulate the immune system and repair damaged cells. Sarah Pitt and her team at the University of Brighton are currently working on creating a new natural antibiotic, made from the slime from the brown garden snail, which they have discovered kills a particular bacteria, *pseudomonas aeruginosa,* which causes blood infections, chronic wound infections, and respiratory infections in people with Cystic Fibrosis. If they manage to successfully create a new antibiotic, they could potentially save many people's lives. It is exciting to see what else the humble snail's slime could be used for in the next few years.

Biologically, snails are incredible too. They are hermaphrodites, having male and female reproductive organs, although they need another snail

---

64 A compound consisting of two or more amino acids linked in a chain

to mate with. They cannot hear, and their eyesight is not very effective, but they have an excellent sense of smell which they use to locate food. They have an amazing number of teeth – 14,000- which are on their tongues. Although they have a simple brain, scientists have discovered that they have long-term memories. They can lift ten times their body weight into a vertical position. It might not come as a surprise that they are also one of the slowest animals on earth. However recently, scientists at the University of Exeter have discovered that snails are speedier than they thought. Dr Dave Hodgson and his team carried out experiments in a garden and discovered that snails can move at one metre per hour. So, they are still not exactly nature's equivalent of Usain Bolt, but they can explore the length of an average British garden in one night. Dr Hodgson also discovered that they move along the trails left by other snails in order to save energy because it takes a lot of energy to produce slime. [65]

Incidentally, there are world championship snail racing. The current champion is a snail called Sammy from Grantham in Lincolnshire who covered the thirteen-inch course in 2 minutes 38 seconds. [66]

Why are freshwater wetland habitats important?

Unbelievably, only 1% of all the water on earth is freshwater. However, humans depend on freshwater for our drinking water, washing, and industry. For the natural world, wetland habitats provide important ecosystems which are linked and maintained by water, allowing trees and plants to grow and a home or breeding ground for many fish, insects, and animals, including some of our rarer wildlife. Forty per cent of all the species on earth rely on wetlands. Wetland habitats can also help to reduce flooding by slowing down the flow of water and giving water somewhere to go. In addition, although worldwide wetlands only cover 6% of the planet, they absorb a third of the carbon dioxide in the atmosphere, making them particularly important at the moment as the world struggles to reduce the effects of climate change.

---

65 The downside of this is that snails spread lungworm which can be fatal if ingested by dogs, so watch your dogs in your garden, especially after wet weather or at night.
66 If you think you have a faster snail the website is www.snailracing.net.

Threats to freshwater wetland areas

Over the last century 90% of wetlands have been destroyed. Our remaining freshwater habitats, unfortunately, at the present time, are in an extremely poor state and this has a huge impact not only on our wildlife but also how we live our lives. In the last fifty years, 83% of freshwater species have declined. Of the 100,000 species which rely on freshwater habitats 13% face extinction.

Traditionally every village in Britain had a pond but, in the last century, about 90% of these ponds were destroyed and of the remaining 470,000 ponds in the United Kingdom about 80% of them are polluted because of lack of maintenance. Of these, most of the polluted ponds are contaminated with nitrogen and phosphates because of agricultural run-off. However, other ponds, as well as rivers, lakes, and streams, have been found to also contain pesticides, heavy metals, sediments and, in the case of rivers, human waste. The Freshwater Habitats Trust asserts that 87% of streams east of a line between the Humber and the Dorset coast are biologically degraded, as well as 95% of canals in England. Worse still British lakes are severely affected by pollution, leading to an increase in algae, which affects the oxygen levels in the lakes, which can reduce biodiversity and affect water quality, and 86% of rivers in England and Wales are in a poor ecological condition. The river system has been damaged by pollution from industries, agricultural run-off, heavy metals from the mining and quarrying industries, and human waste. I recently took part in a campaign to try and prevent water companies from regularly polluting the river system with human waste. Although this is something which should only happen in extreme circumstances, to prevent the sewage systems backing up and polluting homes, an investigation by the Guardian discovered that in 2019 1.5 million hours of untreated sewage was emitted into our river systems on more than 200,000 occasions. More than sixty incidences a year should trigger an investigation by the Environment Agency but, due to lack of funding, the Environment Agency largely allows the privatised water companies to monitor themselves. In 2020 the government told water companies to monitor their sewage overflows, but this has not been done in most cases. The impact of this on human health and the ecosystems in wetland habitats is deeply concerning. Pollution can affect the gender of fish, can be ingested by animals, and get into the food chain. Harmful bacteria can also cause severe illness and death to humans, so people are advised not to swim in rivers and lakes. It costs £1.2 billion per year to remove pollutants from our river systems.

One of the waterways affected by human waste was a chalk stream in Buckinghamshire. Chalk streams are very rare habitats. There are 200 globally and 85% of them are in South and East England. These streams emerge from chalk aquifers so are very pure and rich in nutrients. They are important and biodiverse habitats for some of our rarest species including the white-clawed crayfish, Atlantic salmon, and brook lampreys. Unfortunately, many are in an extremely poor state because of pollution and overfishing.

Failure to maintain our river system also causes an increase in erosion and contributes to flooding. It is worried that climate change will make the problem worse by increasing the temperatures in these habitats, affecting oxygen levels, and causing more floods and droughts. In addition to this our waterways have come under pressure for industries and for domestic water use. Shockingly, a third of the water we extract from our waterways is wasted.

There are three thousand non-native species in the United Kingdom. Most of these are harmless but some of them seriously affect ecosystems. Invasive species such as killer shrimps and signal crayfish are playing havoc with our fragile freshwater ecosystems. Signal crayfish were introduced from North America in the 1970s. Unfortunately, these carry a plague which has had a detrimental effect on our native, white-clawed crayfish population. Worse still the signal crayfish are bigger than our native crayfish, so they outcompete them when it comes to food and habitat. They also bury into riverbanks, causing erosion, and increasing sediments. The invasive signal crayfish are voracious predators, feeding on frogs, fish, invertebrates, and also eating other signal crayfish. They are decimating ecosystems and having a catastrophic effect on biodiversity. Signal crayfish can disperse over 984 feet (300m) in two days, and can cross natural and artificial barriers, and travel overland. It is already well established in rivers and canals in the South-East of England. Signal crayfish is predated by otters, mink, European eels, and Atlantic salmon. People tried to trap signal crayfish, but this had the unintended consequence of allowing more smaller juvenile crayfish to disperse, so now river users are being asked to check for crayfish, and to disinfect rods and boats before using different waterways. It is hoped this will reduce the dispersal of signal crayfish, but the problem is that once an invasive species is established it is extremely difficult, and expensive, to get rid of it. It costs billions of pounds a year to try and control and eradicate invasive non-native species.

Killer shrimp and demon shrimp are also highly invasive species which are in U.K. waters. Originally native to the Black Sea, these shrimps have quickly spready across Western Europe in the last couple of decades, reaching the United Kingdom about a decade ago. Killer shrimps can tolerate a range of environmental conditions but prefer slow-moving water. They breed quickly with one shrimp giving birth to about 450 shrimps per year. They kill a wide range of species for the sake of it, which can have a huge impact on the ecological condition of rivers. Research at the University of Plymouth demonstrated that the mere presence of the killer shrimps affected resident prey behaviour. Killer shrimps can currently be found in Cambridgeshire, the Norfolk Broads, the Midlands, and South Wales. The demon shrimp is the killer shrimp's smaller cousin. Unfortunately, it has spread further than its larger relative. Like the killer shrimp, the demon shrimp decimates ecosystems.

It is not only animal species which can be a problem. Invasive plants can also disturb ecosystems and cost billions to try and eradicate. As well as Himalayan balsam, which dominates wetland areas, leading to a loss in diversity and impeding water flow; there is also floating pennywort which forms dense mats on the surface of water, preventing light getting to other aquatic plants, affecting the amount of oxygen in the waterway, affecting fish populations, damaging ecosystems, and increasing the risk of flooding. As it grows nearly eight inches (20cm) a day it can soon cover a waterway. The only way to deal with it is to regularly cut it back.

Wetland areas are also under threat from big infrastructure projects. The R.S.P. B's Minsmere reserve in Suffolk is one of Europe's most important areas for biodiversity and is home to many of Britain's rare species, including bitterns, avocets, and marsh harriers. However, it is currently under threat from the neighbouring nuclear power station. E.D.F. are going to build a new power station, Sizewell C, which the R.S.P.B. believes will increase pollution, affect water levels, cause erosion, and disturb wildlife. Meanwhile, HS2 continues to destroy nature reserves, ancient woodland, and sites of special scientific interest. It has already destroyed an area the size of Leeds, decimating sites where rare animals, such as the white-clawed crayfish and willow tit resided.

Drainage and modifying rivers, so they do not flow the way they want to, also affects freshwater habitats. For decades conservationists tried to straighten rivers and introduced artificial structures like weirs and dams. This affected the migration of species such as fish, especially Atlantic salmon and brown trout trying to return to their breeding grounds. It has also increased the risk of flooding which has been made worse by floodplains being developed for housing and industry.

Currently avian flu is decimating wildfowl populations, especially in Scotland. Every Winter 40,000 barnacle geese fly from Svalbard in Scandinavia to the Solway Firth. Last Winter 40% of them died as a result of avian flu. Avian flu is easily spread through airborne transmission and faeces. It affects the bird very quickly, causing symptoms such as respiratory problems and a swollen head, and normally causing death in about twenty-four hours. As such it is very hard to contain. It is particularly affecting migratory birds and birds which nest in colonies, such as many seabirds. Although it is rare for humans to catch avian flu, people should not go near dying or dead birds and should keep pets well away from them too.

How are wetland habitats being improved?

Fortunately, many organisations are working hard to improve freshwater wetland habitats. These are just some of the projects taking place.

At Holnicote in Somerset the National Trust are allowing the River Aller to meander naturally through multiple channels and pools. This is the first such project in the United Kingdom. The idea is to naturally restore the landscape, allowing the river to follow its natural course, rather than being forced to go in a straight line like conventional river restoration projects, thereby reducing flooding and helping wildlife. They are hoping to release water voles. This is part of the National Trust's Riverlands project which will cost £14m and improve seven rivers.

Where I live, the North York Moors National Park is improving the River Esk, which is home to several rare species, by planting riparian woodland alongside the river and improving farm infrastructure. It is hoped that the trees will stabilise banks, regulate temperatures, and improve habitats. The National Park also has the RyeVitalise project which is a £3.4 m scheme to restore the River Rye's 257 miles (413 km) square catchment area. The scheme hopes to improve the water quality, slow water flow, and create and connect habitats. They are doing this by planting riparian woodland, creating marshes and grasslands, and installing water improvement measures.

Recently it has been announced that the Rivers Trust, National Trust, and other organisations have announced that they are going to plant over seven thousand acres of new riparian woodland between Devon and Cumbria in order to reduce flooding and improve biodiversity.

In her fascinating book, *Wilding,* Isabella Tree, who along with her husband have transformed their Knepp estate in Sussex from unproductive agricultural land to one of the most important habitats in the United Kingdom, which is home to many rare species, argues that instead of spending lots of money trying to remove water from the land, we should allow the land to naturally return to wetlands and marshland, and rehome people living on floodplains. She gives the example of the Netherlands which is the most densely populated country in Europe with almost eighteen million people living on 16,000 square miles (a sixth of the area of the United Kingdom) and where half of the country is below sea level. Over the centuries dikes were built to try and keep the water out, but in recent years the Dutch government has allowed reclaimed land to return to marshland, wetlands, and rivers and has removed people living in low lying areas and rehomed them on higher ground. Already the Room for the River scheme has reduced the risk of flooding and improved habitats. Isabella Tree makes the point that it costs the British economy an average of £1bn a year to deal with the aftermath of flooding and this is likely to increase as climate change makes severe weather events more common.

She argues that by allowing rivers and wetland areas to re-naturalise flooding will be reduced as well as improving soil and water quality and ecosystems. It is much cheaper to carry out projects which re-naturalise areas, but currently there are no grants to do this.

In this country the Wild Ennerdale project in the Lake District aims," ... to allow the evolution of Ennerdale as a wild valley for the benefit of people, relying more on natural processes to shape landscape and ecology." The project began in 2003 and involved the removal of conifers and concrete fords, bridges, and boundary fences; deciduous woodland has naturally regenerated, sheep have been reduced, and forestry tracks have been abandoned. The result of this is that habitats have improved, helping rare species such as red squirrels and Arctic char. The latter have increased in the lake by 1000%. The area also now has the largest population of the marsh fritillary in England. Mosses, lichens, and juniper trees are thriving. The latter have increased by a whopping 10,000% since the project began, which is particularly welcomed because Ennerdale is Norse for Juniper Valley. The resurgence of natural vegetation is soaking up rainwater and reducing erosion. The devastating floods which did so much damage to the Lake District in 2009 and 2015 did not affect Ennerdale because its soft, absorbent vegetation acted like a sponge, thereby slowing down the water and allowing the River Liza to remain at a normal level.

Near where I live in North Yorkshire the market town of Pickering was regularly flooded, resulting in about £7m of damage. To stop this happening again the Environment Agency, Forestry Commission, and D.E.F.R.A. created 167 natural leaky dams and added natural obstructions, as well as an embankment, and 29 hectares of woodland. Now up to 26 million gallons of water can be stored and then released slowly through culverts. Three months after the scheme was completed it rained for 24 hours on Boxing Day 2015, causing extensive flooding throughout the North of England. However, Pickering's natural scheme worked, and the town has not flooded since the scheme was created.

Elsewhere, a huge new wetland area has recently been created at Wallasea Island in the Thames Estuary. This is because of a partnership with the R.S.P.B and Crossrail. Crossrail have used more than 3000 tonnes of soil which they have excavated from under London, as they have built the new underground rail system, and transported it to Wallasea Island where it has been used to create mudflats, lagoons, and sea marshes, restoring the area to what it would have looked like 400 years ago. They have transformed 740 hectares, an area twice the size of the City of London. Already the nature reserve is an important habitat for migratory and breeding birds, including rare birds like avocets, marsh harriers, brent geese, common terns, and green sandpipers.

The Ouse Fen project in Cambridgeshire is another good example of what can be achieved when industry and conservation charities work together. Hanson are excavating 28 million tonnes of sand and gravel at Needingworth Quarry over a thirty -year period. It is quarrying sections over time. As one section is excavated it is then restored back to nature before being managed by the R.S.P.B. The aim is to create the largest planned nature conservation restoration scheme following sand and gravel extraction in Europe, covering an area larger than 980 football pitches by 2030. Part of this will be the creation of the United Kingdom's largest reed bed which will cover an area equivalent to 644 football pitches.   The project began in 2001 and is already reaping rewards. In 2007 the first bittern appeared, in 2016 there were six pairs of marsh harriers nesting, in 2018 a little crake was heard in the county for the first time in 154 years, and in 2019 a pair of European cranes bred in the area. Water voles, otters, turtle doves, bearded tits, the rare Norfolk hawker dragonfly, and nodding bur-marigold have also been seen.

Meanwhile in Oxfordshire the largest remaining alkaline fen in central England is being restored. This is an internationally important habitat only found in Oxfordshire, East Anglia, and Anglesey. Alkaline fens are very biodiverse and home to many rare plants including grass-of -Parnassus, marsh helleborine, and bog pimpernel. A small pink marshy loving plant has become an unlikely hero of fenland restoration. Scientists in the Netherlands discovered that marsh lousewort is an, "ecosystem engineer." Like yellow rattle it is semi-parasitic, drawing water and nutrients from other plants. It particularly likes reeds and rushes, attaching itself to their roots, restricting their growth and density, thereby allowing other plants to grow. It is also a useful source of nectar and pollen, helping pollinators.

In Wales, the Pen Llyn and Anglesey fens have been restored, in one of Europe's largest fen restoration projects, restoring an area of fenland larger than a thousand football pitches. By removing soil and vegetation and restoring springs and streams, the habitat has been greatly improved. The fens are now home to rare species as well as having better water quality, better carbon sequestration, and reducing flooding.

It is not just wetlands and fenlands that are being restored. Various projects are also cleaning up our river systems. The River Trust has a project called Replenish which aims to improve rivers in London and Northumbria and an ancient fen in Kent. As well as creating four wetlands in the Lee catchment area in Enfield, Waltham Forest, Haringey, and Hackney; the scheme hopes to replenish 1600 million litres of water, capture ten tonnes of carbon per year, improve the biodiversity in 66 hectares, and plant 9000 more trees in Morpeth, Northumberland, to alleviate flooding and pollution. The National Trust's Riverlands project aims to improve habitats and water quality and mitigate the effects of climate change. The project will cover a dozen rivers, including the rare chalk stream – the River Bure in Norfolk-and River Derwent in Cumbria, and will use a variety of techniques including using natural devices to slow water flow, repair riverbanks, and tackle invasive species.

Near where I live, on the River Ouse, at York, there is a conservation project to save the extremely rare tansy beetle, a metallic green small beetle which only lives on tansy plants and only travels 656 feet (200m). Although it has wings it rarely flies. There are only two places in England where this beetle can be found – the River Ouse and a smaller population in East Anglia. In 2016 the York tansy beetle population was found to be 40,000. Since then, more than 400 tansies have been planted. This is an aromatic yellow plant which can be seen on riverbanks between July and October. I saw it quite regularly as a child in the 1980s, but it has become a victim of invasive plants like Himalayan balsam which have taken over riverbanks. This project has had a positive effect on the beetle. Part of the project on the Ouse is to remove Himalayan balsam, reduce grazing, and create safe havens where the beetles can escape from summer floods. Incidentally, tansy is good for the tansy beetle but not for humans. It contains a poisonous chemical called thujone which even if taken in small doses can be fatal.

The Canal and River Trust's Unlocking the Severn Project is reopening 150 miles of the River Severn to make it easier for fish to migrate. In particular, they hope to help the rare twaite shad, a member of the herring family, whose numbers have been reduced dramatically since weirs were constructed on the river during the Industrial Revolution.

In Scotland, the River Till and its tributaries are being restored in order to improve habitats, reduce pollution, sequester carbon, and improve flood management. The partnership is using 'green-engineering' to improve water flow and water storage and remove artificial barriers to help fish migrate. Nature Scotland is also restoring the River Tweed by removing invasive plant species and planting riparian woodland.

Ponds are also being restored. The Freshwater Habitats Trust has the Million Ponds Project a long-term project to recreate the million ponds which have been lost in last century. They want to create an extensive network of new clean- water ponds across the United Kingdom. The Wildfowl Wetlands Trust is restoring farmland ponds and has worked with the Norfolk Ponds project and the Big 50 ponds project in the Severn Valley.

Britain's lakes are also being restored. The Loch Lomond Fisheries Trust Endrick Legacy project aims to tackle invasive species, especially mink and signal crayfish. In the Lake District, the R.S.P.B and its partners have the Futurescapes project which they hope will improve water quality and habitats. The Lake District has some important habitats including Bassenthwaite Lake which has the U.K.'s rarest freshwater fish, the vendace, which had become extinct a few years ago but was recorded in the lake again eight years ago. By working with local farmers to reduce agricultural run-offs and improving the local sewage works, to prevent human waste entering the lake, and removing fish species which were predating the vendace, wildlife in and around the lake is improving, as is the water quality.

What you can do

Create a pond in your garden. There is lots of help and advice on the interest especially on the Freshwater Habitats Trust site (www.

freshwaterhabitats.org.uk), R.S.P.B (www.rspb.org.uk) and the Wildlife Trusts (www.wildlifetrusts.org.)

If you are a farmer as well as creating a pond on your land also have a look at ways of storing slurry and preventing agricultural run-off. If you approach the conservation charities above or charities such as the River Trust, advice is available.

You could become a volunteer for one of the conservation charities (information is at the back of the book.) Tasks like removing Himalayan balsam are labour intensive but vital for restoring wetland habitats. There are also, as I hope I have demonstrated, lots of ongoing projects throughout Britain, all of which require volunteers.

If you are a regular water user conservation charities, such as the River Trust, and Canal and River Trust, ask you to Check, Clean and Dry to prevent the spread of invasive species. More details are on their websites- www.therivertrust.org and www.canalrivertrust.org.uk.

Think about how much water you use (150 litres per day on average) and about what you put down the toilet (basically it should just be pee and poo) and think about how you can save water. For instance, have a water butt in the garden to catch rainwater rather than using a hose to water your garden. Apparently putting a bottle of water in your toilet's cistern can save water. Maybe just wash up once a day. Showers use less water than baths. Do not pour oil and chemicals down the sink or toilet. There are many natural products available. Ofwat has various tips at www.ofwat.gov.uk and Friends of the Earth at www.friendsoftheearth.uk/ 13 ways to save water.

The deep blue sea

The British Isles has approximately 11,073 miles (17,819km)[67] of coastline. Scotland and its islands make up 60% of the coastline. Great Britain is made up of over a thousand islands[68], with 219 of them being inhabited. Coton-in-the-Elms in Derbyshire is the place in Britain which is furthest away from the sea, and that is only seventy miles from the coast. As a nation we have a special affinity with the sea.

Of course, coastal habitats cover a wide range of habitats, from the sea itself to the coastline to maritime cliffs and saltmarshes to sand dunes and beaches.

The sea is affected by phenomena such as ocean currents, which influence temperature, and, of course, tides. It is affected by the gravitational pull of the moon and the sun, but as the moon is closer to earth than the sun its gravitational pull is twice as strong as the sun's. This gravitational pull keeps the earth orbiting around the moon, but it is stronger on the side of the earth which is closest to the moon. On this side, the moon pulls everything which is free towards it, whereas on the far side of the earth things move away from the moon. The nearside of the earth has a bulge towards the moon, and the far side of the earth has a bulge away from the moon. As the earth spins round different parts of the earth move underneath these bulges causing high tides. In Britain we have two tides a day. The interaction between the solar and lunar gravitational pull creates spring and neap tides. The former are tides with the highest difference between high and low tide. These happen every full moon. Neap tides occur when the moon is in its first or third quarter.

Beaches and estuaries are always changing because of the daily tides.

---

[67] According to the Ordnance Survey. However, this figure is contentious, and some people estimate it as being more like 7660 miles, whereas others estimate it is 19,491 miles. It seems that it is all depends on how you do the calculations and what you count as the British Isles and what you count as an island.

[68] Again, figures vary massively depending on the definition of an island.

The Wonders of the Wild Places

However, there are many interesting creatures which can be found on our beaches, especially in rockpools. Have a look next time you go to the seaside. The best place to look is in rockpools nearest the sea in late Spring or early Autumn.[69] You might see a starfish, which are found in every ocean in the world. There are more than two thousand different species. They do not have a brain, heart, or blood but they do have a specialised stomach which they can extend out of their body and digest food, which is too big to fit in their mouth, then suck the food into them. Most starfish have five arms, but they can have up to 24. They have a good trick if they get attacked by a predator – they can regrow an arm- not only that, they can regrow a new body from a lost arm, although they have to store up nutrients first.

Also look out for sea anemones which are actually animals rather than plants. They are related to jellyfish and coral. Like starfish they do not have a brain, heart, or blood but do have stinging tentacles which they use to catch plankton, microscopic creatures which provide food for a lot of sea-dwelling animals. The most popular sea anemone is called the beadlet anemone which retracts its tentacles between tides to keep them moist.

If you are lucky you might also see one of the 950 species of sea urchins. Urchin is an Old English word for hedgehog, and like hedgehogs, sea urchins are a prickly animal. They can live at depths of 16,000 feet (4.8km), eating algae, but if you do come across one in a rockpool or in the sea do not touch it. They can be very venomous.

More common things which you are likely to see in rockpools are limpets and crabs. Limpets are actually a name given to several aquatic snails which are not really related to one another. Incredibly, they have more than a hundred teeth but only the outmost ten teeth are used for feeding. Unsurprisingly, their teeth wear out very quickly and are replaced every 12-48 hours. They feed on algae by scraping it off rocks with their teeth, which are the strongest biological material in the world. Scientists at the University of Portsmouth are looking into whether they can copy the structure of limpet teeth and use it to make parts for boats, aeroplanes and Formula 1 racing cars. Limpets can live up to twenty years. The most

---

[69] Remember to check the tides. I have been caught out a few times.

common ones you will find in British rockpools are the common, black-footed, and China limpets. The latter can just be seen on beaches in the South of England. Crabs have been around for about 200 million years and there are about 4500 species worldwide ranging from the 12 ft (3.7m) Japanese spider crab to the half-inch (13mm) pea crab. You are most likely to see crabs in rockpools or walking sideways across the beach. The most seen one in rockpools is the shore crab which is greenish. Crabs can regrow their shells, except the hermit crab which just steals other crab's shells. A crab normally moults and then regrows their shell a couple of times a year. Their shells are actually an external skeleton. They have eyes on stalks and can look in different directions at the same time, looking for predators. They have a lot of copper in their blood, so their blood is blue. Scientists have recently discovered that not only do crabs feel pain, but they also remember it. Crabs are covered in tiny hairs called setae which they use to detect chemicals, touch, and movement. The word cancer is derived from the Latin word for crab. Ancient Greek doctors Hippocrates and Galen noted the swollen tumours in crab's veins.

You might notice barnacles attached to a rock, boat, or any other hard surface. These crustaceans are related to crabs and shrimps. They are in two broad forms – tiny conical-shaped barnacles and goose barnacles (for centuries barnacle geese were thought to come from these barnacles!) which makes shells out of several pieces. These creatures, which can live up to a dozen years, do not have particularly interesting lives. They start life as planktonic larva before becoming an adult, at which point they find a hard surface and bash their head against it, breaking their glue glands, and gluing themselves head- first to the surface for the rest of their lives. They eat through their legs, waving them around to catch plankton when the tide comes in.

Something which is regularly seen in rockpools and on beaches, as well as in the sea, is seaweed. This is actually not a plant, but a multi-cellular algae. In fact, it is the common name given to several hundred species of marine algae. Some seaweed is microscopic such, as phytoplankton. Plankton is derived from the Greek word meaning *drifter or wanderer* which describes it perfectly because these marine drifters are poor swimmers, carried by the tides and currents. Some spend their entire life cycles drifting, whereas others just drift when they are young. Most plankton are microscopic, but some crustaceans and jellyfish also fall within this classification. There are many classifications of plankton depending on size, type, and how long they drift, but the main classifications are phytoplankton (plants) and zooplankton (animals.) Phytoplankton lives on the surface of the oceans because they need to photosynthesize. Plankton is described by the National Geographic Society as the "...hidden heroes ..." of marine ecosystems providing food for a number of species. Phytoplankton is vital for the survival of all living things on earth, including humans, because it produces between 50-80% of the world's oxygen (depending on which source you read) and absorbs carbon dioxide. It requires nutrients such as nitrogen, phosphorous, and calcium to thrive.

Zooplankton, such as sea snails, pelagic worms, and krill, spend daylight hours lurking in the deep, trying to avoid predators; then, during the night, they come to the surface of the oceans, where they feed on the phytoplankton. This is the largest migration on earth. It is so big it can be observed from space.

There are also bacterioplankton, which degrades organic material, and viroplankton which carry viruses.

Plankton is an important part of the marine ecosystem[70] but is very sensitive to changes including temperature, salinity, pH levels, and nutrient concentration in the water. Many types of phytoplankton can grow and reach high numbers when nitrogen and phosphorous are added by human activities. This causes harmful algae bloom which releases toxic chemicals which kill fish and filter feeding organisms, such as oysters, which if consumed by humans can cause illness or death. Although these harmful algae blooms have existed for centuries they are becoming more frequent, and are happening in greater numbers in recent years, which has a knock-on effect on the zooplankton and bacterioplankton. The effect is mitigated by zooplankton-grazers who have varying susceptibility to harmful algae bloom toxins. Another consequence of these harmful blooms is a reduction of oxygen at night which affects other marine species.

---

70 Plankton also occurs in freshwater. Clear water has less plankton than green or brown water.

The Wonders of the Wild Places

There are many different species of seaweed spanning various colours, but mainly brown, green, and red. Although they are not plants they do need to photosynthesize in order to obtain energy, so they normally live at the edges of oceans. Most varieties of seaweed must be attached to something. One of the most popular varieties on British beaches is kelp which has root-like 'holdfasts' which they use to attach firmly to rocks. Kelp can grow two feet (61cm) per day. Seaweed absorbs nutrients and water through its surface tissue. It is an important food source for marine animals, and can also be eaten by humans, and used as an ingredient in products such as toothpaste, paint, and cosmetics.

Seagrass is the only flowering plant which can live in seawater. There are four species in the U.K. – two species of tasselweeds, and two species of eelgrass. Although it lives under water it still needs to photosynthesize so it can only be found in shallow water (up to a depth of 13ft (4m.) Seagrass are described as the, "lungs of the ocean," because they absorb carbon dioxide and release oxygen. A metre square of seagrass generates ten litres of oxygen per day. Their roots are anchored in mud or sand, so they also help stabilise the seabed, thereby reducing erosion.

Sadly, due to alien species, pollution, and dredging globally 11,583 square miles (30,000 km^2) of seagrass has been lost this century.

Beaches are defined as a strip of land, made up of loose sediment (normally sand or stone) next to a big body of water. A third of the United Kingdom's coastline are beaches, with its shingle beaches, which tend to be found mainly in the South- East of England, being of particular global significance because they are only found in a few places in the world, and the United Kingdom has 6000 of them. Dungeness in Kent is the largest shingle beach in England, made up of Ice Age flint. It has 2000 hectares of shingle. There are only five other shingles beaches over a hundred hectares in the United Kingdom, so these are extremely rare habitats.

Shingle beaches are important habitats because they act like deserts. The pebbles warm up during the day and cool rapidly at night. They are dry habitats. Only rare, specialised vegetation, such as orache, sea holly, sea pea, and yellow-horned poppy can live in these conditions. Vegetated shingle is an internationally rare habitat but can be seen at places such as Dungeness and Rye Harbour. There are four zones of vegetation –

ephemeral (just above the tide line), pioneer plants, which bind the shingle together, intermediate (which has less diversity) and then most landward zone with plants like bittersweet, sea bindweed, sea campion, and sea holly. Orache looks like a green haze. It lives just above the tidemark but still gets washed away by Autumnal storms. Sea pea has an extensive root system which allows it to grow like a mat across the shingle. This species is particularly rare and can only be found at Winchelsea and Rye in Sussex. Yellow- horned poppies get their name from the horn-shaped seed pods which are the largest seedpod of any wild British plant. It has long roots and hairy leaves which help it reduce water loss. Sea holly is low-growing and has distinctive silvery leathery leaves. In the 17th and 18th centuries its roots were harvested and turned into lozenges. It is used as a diuretic and used to treat conditions such as cystitis, kidney stones, and inflammation of the prostate. [71] Sea campion has an extensive root system, useful for binding shingle together. It is also known as dead man's bells or witches' thimbles. It was thought that if someone picked it they would die. As it also grows near cliffs there might be some truth in it! The name campion supposedly derives from the Latin *campus,* which originally referred to sports fields. The flowers were included in garlands awarded to winners of Roman games. The word *champion* also derives from the same root. Sea kale also grows in the pioneer zone. It is a type of cabbage with waxy leaves which reduce water loss.

Shingle beaches are also important habitats for invertebrates and migratory birds, especially common terns which like to lay their eggs on the shingle. They have a tail a bit like a swallow, with long 'streamers', so are quite easy to spot. Invertebrate species such as the short-haired bumble bee, and brown-banded carder bee, like shingle beeches, as do rarer species, including the whelk-shell jumping spider, which as the name suggests hides in whelk-shells. Short-haired bumblebees used to be widespread below the Humber. However, they became extinct in 2000. Swedish Queen bees were reintroduced at Dungeness in 2011, and along with another rare bee, the ruderal bumblebee, which has also been recorded at Dungeness, they are thriving locally and have been seen in

---

71 Always seek advice from a trained herbalist.

some areas where they have not been recorded for a quarter of a century.

Even if you do not live near a shingle beach there is a lot to see on sandy beaches too. You might find mermaid's purses which are actually the eggs of a dogfish, a member of the shark family.

Some species are too small for you to see, but worth mentioning because they are amazing creatures. One such is the sadly rare water bears. These miniscule creatures, only 0.1mm-1.2mm, are said to look like a bear if looked at under a microscope, but I do not see the resemblance myself. Their other name are tardigrades which means slow paced. They live in freshwater and seawater as well as moss and beach sediment, using their straw -like mouths to feed on plant sap, although some seawater species are carnivorous, eating other tiny animals called rotifers and nematodes. These microscopic creatures might be small, but they can not only withstand extreme pressure, but also temperatures ranging from -272°C to 151°C. They have the ability to also perform a process called cryptobiosis which means that they can dry out and exist without water for several years, but then a small amount of water can rehydrate them, and they can come back to life in a matter of minutes. In 2007 Ingermar Jonsson at Kirstianstad University in Sweden sent two dried out species of tardigrades into space. Miraculously most of them survived ten days in space, and even rehydrated when they returned to Earth. They have also endured exposure to lethal levels of ultraviolet light, just absorbing the light and emitting it as a harmless blue light. Unsurprisingly, they have survived all the mass extinctions on Earth.

Another small species which you might see on the beach at night, eating the seaweed, are sandhoppers, also known as sand fleas, which can jump several inches. These tiny creatures are an important food source for many species of fish and birds.

One of my favourite things to do on the beach is to look for seashells. These are an important part of a beach with the organisms in the shells providing food for many species, and the shells breaking down over time to form part of the beach.

Most seashells come from molluscs which comprise of over 11,000 species and are the largest phyla of animals behind insects and worms. It is a diverse group including limpets and oysters, snails, and slugs, but also octopus. Molluscs use calcium carbonate and proteins to build their shells, and aragonite to build the inside of their shells. The aragonite helps to repair the shell if it becomes damaged. The protein prevents the shell dissolving in the seawater. As the molluscs grow so does the shell. If you look carefully you might notice the growth rings, which are like tree rings, and help scientists to study things such as climate change.

There are five main broad types of molluscs. Many gastropods live in spiral shells, although slugs also fall into this category. Some gastropods are scavengers and predators. Some of them have a siphon to move water into their shell in order that they can determine if prey is close by. Others have a sharp radula which they use to eat seaweed. Bivalves include clams and oysters. They have two parts to their shells. They are mainly filter feeders. The water passes through the shell and across membranes (cilia) which trap any plankton, which they eat. A single oyster can filter fifty gallons of water per day. That is the equivalent of 36 Olympic swimming pools. Oysters change gender at least once during their lifetime. They are sensitive to water quality, especially if the water is low in oxygen, and need clean water in which to survive. Polyplacophora have eight protective plates and live in the water. Scaphopods are tusk shaped with an opening at both ends. Monoplacophora were thought to be extinct but live ones were discovered in 1952. They have been around for about 500 million years. They have remarkably simple shells and eat organic debris from the ocean floor, using a mouth on the underside of their bodies.

Ninety per cent of shells open to the right so it is exceedingly difficult for molluscs with left opening shells to mate.

Many seashells are a helix, a type of smooth space curve with tangent lines at a constant angle to a fixed axis. One of the remarkable things about the natural world is how often these helices occur. As well as seashells, snail shells, and horns they also occur in D.N.A, weather patterns (such as hurricanes and cyclones), the way stems appear on a plant   (phyllotaxis), algae and ferns, fingerprints, and galaxies, as well as many other things. Even the way in which a bird of prey approaches its prey and the way an insect approaches a light source follow the same principle of parsimony which is a scientific principle which says that things are connected, or behave, in the simplest, most economical way. Thus, plants tend to use spirals so that all their leaves have maximum exposure to the light.

Fibonacci numbers also appear throughout the natural world. This is a sequence of numbers where each number is the sum of the previous two, so the sequence starts 0,1,1,2,3,5,8,13,21,34, 55, 89... They were discovered by an Italian mathematician called Leonardo Pisano who was born about 1170 in Pisa. One of his manuscripts was the Liber Abaci which referred to Indo-Arabic numbers, which gradually replaced Roman numerals. In his manuscript he posed many questions, one of which was, "How many pairs of rabbits will be born in a year, starting with a single pair, if each pair give birth to a new pair which becomes reproductive from the second month?" If, like me, your head explodes when posed with a mathematical problem, fortunately Pisano gave the answer, which, as it turns out, is the Fibonacci sequence, although it was not called that until 1877 when a mathematician called Edouard Lucas published a number of important papers on the sequence. For the purposes of this book the important part of the Fibonacci sequence is that the value between any number and the previous one is always the same number – 1.618...This is known as the golden ratio and occurs throughout nature because it creates a round shape with no gaps, such as a sunflower or a snail's shell.

Sand dunes are formed when sand and seashell fragments are blown until they get trapped amongst the plants growing along the strandline. Over time these dunes increase in size, becoming as high as sixty feet tall. Some of the sand dunes in the United Kingdom are 9000 years old. Coastal grasses such as marram grass and pamprass grass have deep roots which help bind the sand together. Marram grass also helps other plants, like sea holly and sea bindweed, to grow. Both grasses are tolerant to high winds, drought, and sea spray. Another plant which helps the formation of sand dunes, by trapping windblown sand, is sea rocket, whose buoyant seedpods are carried by sea currents and then germinate along the strandline. Look out for its mass of lilac flowers which are extremely popular with coastal pollinators.

Some of the best sand dunes in the United Kingdom can be found at Lindisfarne, off the coast of Northumberland, Kenfig nature reserve in Glamorgan, Holkham nature reserve in Norfolk, Forvie near Aberdeen, East Head in Sussex, and Portstewart, near Londonderry in Northern

Ireland. These places, and other sand dunes, provide an important habitat for rare species such as natterjack toads, sand lizards, rare orchids, insects, and fungi.

Sand lizards are extremely rare in the United Kingdom only existing in three small pockets of the South of England, and the North and West coasts of Wales. In the spring the males have green flanks (known as badges) and bob up and down trying to look bigger. The males with the brightest flanks are the ones the females mate with. In the winter they dig tunnels into the dunes in which they hibernate. In the spring the female lays her eggs in the dunes, hoping the sun will help to incubate them.

Regarding rare orchids, at Kenfig in Glamorgan you can see Autumn lady's- tresses, the last native orchid to flower. It has six-inch (15cm) spikes of white florets that are twisted like braided hair, hence the name. Pyramidal orchids have a densely packed pyramid of pink flowers and are often found near marram grass. They are a popular source of pollen for long-tongued moths.

Other plants to look out for on sand dunes are kidney vetch, scarlet pimpernel, and viper's bugloss. Viper's bugloss grows in a variety of habitats including heaths, coasts, shingle, and sand dunes. These tall purple flowers are popular with pollinators. They get their name from the shape of the flower, which looks like a coiled snake when in bud. Scarlet pimpernel might sound like a French spy, but its name comes from the old French word *pimper* which means trim or smart. It is a good name for this scarlet, perky plant which spreads along the ground. On sunny days it opens its flowers at eight o'clock in the morning and shuts them again at three in the afternoon, hence some of its vernacular names, like shepherd's sundial. It was believed to be a magical flower, maybe because it was said to grow at the foot of the Cross and the flowers were made out of drops of Christ's blood. Anyone holding scarlet pimpernel was said to have second sight. Medieval herbalists used the leaves of the plant in order to draw out arrows and splinters from wounds. Kidney vetch is a yellow-flowered member of the pea family, which was used for circulation problems and as a poultice on wounds.

There are numerous insects which live in and near sand dunes. One of these is the sand wasp which paralyses caterpillars before carrying it back to its hole in the sand dunes where it feeds the unfortunate caterpillar to its larvae. Once the sand wasp is fully stocked with enough prey it seals up the entrance to the hole with mud. There are various species of burnet moths. The six-spot burnet can be seen during the day. The caterpillars live on bird's foot trefoil, whereas the adult moths prefer knapweed and thistles. These moths are very distinctive, black with six red spots on each forewing. The other five species of burnet moth in the United Kingdom have five spots. The spots are not just for decoration, they also warn predators that they are poisonous. If under attack the caterpillars or moths release hydrogen cyanide. This makes them taste unpleasant but also, in sufficient quantity, can kill predators. The female also releases cyanide to help attract males during the mating season.

The equally brightly coloured, red, and black, cinnabar moth flies during the day and at night. It lives on the common ragwort, a yellow plant popular with pollinators but unpopular with farmers and horse riders because it can be toxic if ingested by cattle and horses. Ragwort has been in Britain since the Ice Age and, as well as providing food for the cinnabar moth, it also provides a habitat for seven species of beetle, twelve species of flies, and seven other moths which rely solely on ragwort. As it flowers in the later summer when many other sources of nectar have died back it is also an especially important nectar source for at least thirty solitary bees, eighteen species of solitary wasps, and fifty insect parasites. In total 177 species use ragwort as a source of pollen and nectar, including forty species who use it as a nocturnal source of nectar and pollen. Despite the importance of this plant though, in recent years, there has been a war against it, and a lot of money is spend removing it. This has a huge impact on the insects which rely on it. In the opinion of Isabella Tree, in her book *Wilding*, this is a waste of money and detrimental to the ecosystems reliant on this plant. Ragwort seeds can exist in the soil for a decade before even such a thing as a rabbit scratching the earth can allow enough light to enable the seed to germinate. Also, as Isabella Tree points out, the concern surrounding ragwort is based on a misinterpretation of the science. In 2002 the British Horse Society claimed that 6500 horses a year died from ingesting ragwort, but in reality, most animals know to avoid this plant because it gives off an awful smell. Some horses and cattle do die when the ragwort is part of the silage that they eat, and they are unable to detect its bitter taste. However, even then, according to Isabella Tree, they need to eat about 5-25% of their body weight. She also points out that many plants are poisonous to animals, but animals are aware of it and avoid them. Isabella Tree makes the point that the attitude to ragwort highlights the problems with British conservation, namely deeming that some species are good and worth preserving whilst vilifying other species.

Sand dunes are also an important habitat for some birds. Look out for little terns and sandwich terns as well as ringed plovers near sand dunes. As the name suggests little terns are the smallest of the tern species. Its bill is yellow with a distinctive black tip. These rare, and declining, terns only weigh the same as a tennis ball. They migrate from the Gambia in West Africa to the United Kingdom to breed, a distance of nearly 3107 miles (5000km). In Norfolk they are known as *little picker* because they pick sand eels and herring from the sea with their beaks. One of the reasons for their decline is that they do not choose the best place to nest, often laying up to three very well camouflaged eggs near the high-water mark so they often get washed away in stormy weather, and increasing problem because of climate change, or disturbed by people and dogs. Their chicks are also easy prey for a variety of predators including crows, snakes, and herons. Sandwich terns are white with a black cap and a black bill with a yellow tip. They get their name from the town in Kent where they were first spotted rather than the bread-based snack. They are gregarious birds, nesting in large colonies around the British coast, although mainly in the South and East of England. By the end of the 19th century, they had almost become extinct because of the Victorian's love of egg collecting and the effects of the millinery trade. However, despite their population naturally fluctuating, they are more common than the little terns, but there are only about 11,000 breeding pairs in the United Kingdom.

Another tern found in Scotland, Northern Wales, and Ireland is the Arctic tern which has the reputedly the longest migration of any bird (although some experts give this accolade to the sooty shearwater which is another long-distance migratory bird). Every year the whole family flies from the Arctic to the Antarctic and back. That is a distance of 44,000 miles (70,811km). They have long, powerful wings to help them fly, even in strong winds.

Ringed plovers are small waders which nest around the coast as well as reservoirs and flooded gravel pits. Unlike its cousin, the little ringed plover, it is resident in the United Kingdom, but is joined with ringed plovers from Greenland and Canada in the winter. It eats invertebrates and crustaceans, tempting its prey to the surface by doing a tap-dance to imitate rain.

Britain has some of the most important seabird colonies in the world and is home to 25 species of seabirds[72] including gannets, fulmars, razorbills and guillemots, kittiwakes, Manx shearwater, gulls, puffins, cormorants, and the schoolboy's favourite, shags. Seabirds have been around for 100 million years but there has been a sharp decline of 60% in the last sixty years, with about a third of seabirds facing extinction, mainly due to climate change causing the seas to be more acidic and affecting the temperature of the sea and oceanographic patterns; overfishing, and pollution.

Just under three quarters of the world's population of gannets live around the British coastline, although they do not breed any further south than the Yorkshire coast, with Flamborough Head near Bridlington being a good place to see them. Bass Rock off the Eastern coast of Scotland has the largest colony in the world with about 150,000 gannets breeding on the small island.

Most seabirds live in colonies. Breeding beak by beak with their neighbours has its disadvantages including the risk of disease, competition for food, being cuckolded by one of your neighbours, and even cannibalism, but, in recent years, scientists have realised that there are many advantages too. Not only is there safety in numbers against predators, but scientists have concluded that the colony is an information hub where birds can find out information such as where the fish are. It has become clear that every colony of gannets have their own distinctive fishing grounds, with some flying 350 (563km) miles or more out to sea to catch fish. As Adam Nicholson explains in his fascinating book *A Seabird's Cry*, each colony has a "...pattern of understanding and a way of life tied with its own geography and unique to that gathering of gannets... inherited across the generations."

---

72 18 in Britain and 8 in Ireland

The Wonders of the Wild Places

Gannets may live communally, but they fly individually to their colony's unique fishing grounds. They have incredible eyesight. They can see another gannet twenty miles (32km) away and they have been tracked flying so high they can see beyond the curvature of the earth. Whilst in flight the cornea of the eye fixes the image on the retina, allowing the gannet to see the mackerel, herring, or sand eels underneath the water. When they see fish, they have a habit of literally dropping out of the air, their binocular vision calculating the exact point where they must fold back their wings before they enter the water. They drop from a height of about 90 feet (27m) at about 50 mph, hitting the water at 80 feet (24m) per second. They have two distinctive dives – a V-shaped exploratory dive to check that the fish are there, and then a U-shaped dive to catch their prey. Their face, neck, stomach and back have air sacs which absorb the impact. They have internal nostrils, so they do not inhale water. Australian scientists have discovered that within about an eighth of a second of entering the water the gannet changes its eyes, squeezing and distorting the lenses and opening their irises so that they can see clearly underwater. Gannets have been recorded being killed in underwater collisions as they change their eyesight to underwater mode.

Gannets are also notoriously aggressive with razor sharp bills. Never approach one, and if you find an injured one do not pick it up. In *A Seabird's Cry* Adam Nicholson tells an apocryphal tale of a man trying to help an injured gannet which rewarded his kindness by pecking his eye out.

Gannets are bucking the trend and their numbers have increased in the last few decades. The ones which feed in the North Atlantic are doing particularly well. By contrast another seafaring adventurer, the fulmar, is in decline, possibly because of overfishing. It gets its name from the Old Norse word meaning *foul gull* probably because it can excrete a sticky, smelly oil which it uses to feed its chicks, and to ward off predators because if the oil lands on a predator's wings it makes them incapable of flight. The University of Aberdeen has been studying fulmars since the 1950s. In 2012 they fitted a geolocator to a fulmar with the catchy name 1568. They tracked where he went. They were amazed by the results. In his first couple of days, he had travelled almost a thousand miles (1609km). They expected him to return to land but he kept going, flying to the mid-

The Wonders of the Wild Places

Atlantic, covering 3900 miles (6276km) in a fortnight. The team concluded that he remembered the journeys he took. One of the things he might have remembered are the wind conditions. Fulmars are gliders. Two Scottish ornithologists, Bob Furness and David Bryant, have worked out the amount of energy a fulmar needs to expend in different conditions. If it is calm, they use up 2000 calories in flight (more than thirteen times the amount when they are stationary). If the wind is about 20 mph (Force 4-5) they can glide easily without having to use up very much energy although they do need to beat their wings every second. However, if the wind is 25 mph (Force 5-6) they can glide without hardly every beating their wings and expend virtually no energy.

Guillemots and razorbills look a bit like penguins, with black backs (or very dark brown in the case of guillemots) and white chests. Razorbills are slightly smaller than guillemots, with smaller wings and a longer tail. The guillemot has a stubbier bill. The razorbill's beak has a distinctive white stripe at the tip. The two species often nest together although the razorbills prefer the lower ledges of the cliffs and deep ravines. Razorbills feed on fish, mainly sand eels, herrings, and sprats, which it catches by diving under the surface of the water. Although it normally catches fish in the top 65 feet (20m) of the sea it has been observed swimming a long way under the sea. The fathers take the chicks out to sea with them to try and keep them safe. One of the reasons why razorbills are in decline is that the survival rate for chicks is not good, and the fish they eat, especially sand eels, are in decline. Like razorbills, guillemots also dive many hundreds of feet underneath the sea. Unlike human bones and the bones of many land-based birds, the guillemot's bones are solid, which partly acts as ballast when they dive. Like gannets, guillemots have large eyes which enable them to see long distances. Scientists have also discovered that they are cognitively extraordinarily complex, being able to know and remember information, such as returning to the precise place where they caught fish and the precise breeding sites. Incidentally, according to the Collins dictionary, the word guillemot is a derivative of the French equivalent of William, Guillaume.

Razorbills and guillemots are auks. In 1855 two fishermen on Eldey island, off the coast of Iceland, killed two birds which looked like a bigger version of the razorbill, standing at 2ft 7 (79cm) tall. As they chased and slaughtered the birds one of the fishermen squashed the egg that the female bird had been incubating. In doing so he had caused a once plentiful flightless bird to become extinct. The story of the great auk is a cautionary tale about the disregard that humans show to the natural world. We are the worst predators on this planet. Great auks had lived for a long time and had been revered by Neanderthals and Native American tribes. Thirty-five-thousand-year-old cave paintings of auks have been found in El Pendo cave in Camargo in Spain and in Paglicci in Italy. Great auks looked like penguins, although the two species were not related. Some sources claim that penguins got their names because sailors in the Southern Hemisphere thought they resembled great auks and called them by the great auk's genus name of Pinguinus.[73] However, great auks only lived on a handful of islands in the Northern Hemisphere including St Kilda's off the coast of Scotland, where the last great auk in Britain was condemned as a "...maelstrom-conjuring witch..." in 1844 and beaten to death. Another island it inhabited was Funk Island off the coast of Newfoundland. This island was a popular stopping off point for transatlantic sailors. In 1534 a French explorer called Jacques Cartier claimed that every ship would herd hundreds of great auks onto the ship and eat some fresh whilst salting, "...5 or 6 barrels of them." Despite this a sailor in 1718 described seeing so many great auks breeding on Funk Island that, "... a man could not put his boot between them." Another sailor, Captain Richard Whitbourne, claimed that sailors harvested the auks," ...by hundreds at a time as if God had made the innocency(sic) of so poor a creature to become an admirable instrument for the sustentation of Man." The great auk, which had existed for millennia did not stand a chance. As if being killed for food was not bad enough, things got worse for the great auk because hunters had exhausted the supply of eider down by 1760, because of overhunting, and started to hunt great auks for their down too. By 1810 the colony of hundreds of thousands of birds had become extinct. Every bird had been killed. Despite the British government making it illegal to kill great auks,

---

73 Other dictionaries think penguins are called after the Welsh words pen, meaning head, and gwyn, meaning white. Other sources think it is from the Latin pinguis meaning fat or oil.

at the end of the 18th century the persecution continued with Victorian egg collectors being able to get as much as a year's wages for one egg.

Since birds first flew 140 million years ago there have been about 150,000 species of birds. Unfortunately, 140,000 of these are now extinct and this number is likely to increase if we do not stop and take stock of what human beings are doing to the natural world. It is possible to live harmoniously with the natural world, but we must stop thinking, like Captain Whitbourne, that the natural world is something for us to exploit and plunder and remember that we are part of the natural world and are reliant on it for our future too.

Another iconic bird which might become one of the ever-increasing list of extinct birds is the puffin. In the last twenty years their numbers have declined due to climate change which is affecting their food supplies, as well as overfishing and pollution. Puffins are also hunted and eaten by Icelandic people which contributes to their decline. Since 2015 they have been listed as 'vulnerable' by the International Union of Conservation of Nature which means that they are at a high risk of extinction. Puffins mainly eat sand eels, although they will also catch herring, sprats, and capelin. So far, the temperature of the seas around Iceland have just risen by 1°C but this has had a disastrous effect on sand eel populations, which have had a knock-on effect on seabird populations. This means that puffins are having to travel further looking for food- up to 300 miles (483km) per fishing trip in a bad year. This is difficult for the puffin. Unlike fulmars, puffins are not gliders, and it takes a lot of energy for them to fly.

Puffins come to the United Kingdom every Spring to breed, mainly to the Scottish, Irish, and Welsh coasts, although they also have colonies in Devon and Cornwall, and Bempton cliffs near Bridlington, and the Farne Island and Coquet Island off the coast of Northumberland. The rest of the year they can be found in sea around Scandinavia and Newfoundland, with a few heading to warmer climes in the Bay of Biscay. Scientists have discovered that puffins have light receptors in their brains which helps them trigger when to fly to their breeding sites as the amount of daylight increases. These receptors also trigger a physical change in the puffin, causing it to put on weight and triggering hormonal changes, turning their distinctive beaks from grey to orange to attract a mate. The orange comes from carotenoids from the fish that they eat. They must eat a lot of fish to get enough carotenoids which they need to turn their beaks and feet orange. However, they do not just change colour to attract to a mate, the carotenoids are also antioxidants and regulate the bird's metabolism and immune system so the more orange the beak, the healthier the puffin is. Their beaks get more orange every year. When they reach their breeding colony, they dig a burrow with their beak or feet, or use an old rabbit burrow, and then lay an egg in about mid- May.

During the breeding season, the puffin will spend about seven hours a day out hunting. They normally only dive to about 50 feet (15m), using their feet as rudders, but have been recorded at 220 feet (67m). A puffin must catch 450 sand eels a day to feed its chick (known as a puffling) and this is getting increasingly difficult. It also has a huge metabolic effect on the puffin, the equivalent to a human spending a day knocking down walls.

After the breeding season the puffin fly off on their own. Research by Oxford University has shown that every puffin has its complex pattern of movements, unique to it, which they repeat every year. Each year they return to the same breeding site where they breed with the same mate. Puffins are loyal partners, with some being recorded as mating with the same puffin for twenty years.

Puffins were originally called sea parrots, probably because of their beaks, the word puffin originally referred to a young Manx shearwater, whose Latin name is *puffinus puffinus*. Then in the 17th century, the name was transferred to puffins, with Manx shearwaters keeping their Latin names. The puffin's Latin name is *fratercula* which means friar. They probably get this name from their habit of clasping their feet together and looking as if they are in prayer.

Manx shearwater are one of the five shearwater species seen in the British Isles along with sooty shearwater, Cory's shearwater, great shearwater, and the Balearic shearwater being occasionally seen off the British coastline. The Manx shearwater is the only one which breeds in the British Isles mainly, as the name suggests, on the Isle of Man, as well as the Scottish Isles, Anglesey, the Pembrokeshire coast in Wales, and the Irish coastline. Like the puffin it lays one egg in a burrow and then the chick is fed at night on pre-digested fish. They feed the chick for a couple of months before leaving it to its own devices and heading back out to sea.

Shearwater are related to fulmars, petrels, and albatrosses. In his captivating book, *A Seabird's Cry*, Adam Nicholson describes this family of seabirds as, "...emperors of the wind, the greatest of all travellers..." because these birds embark on huge migrations, annually flying from one hemisphere to the other. A sooty shearwater migrates approximately 40,000 miles (64374km.) Like pigeons, the shearwaters use the sun as a compass, taking into account the position of the sun depending on the time of day. They can tell the time, relate their position to the sun, calculate where the sun is going to be at midday, and work out which way is south. Geoffrey Matthews, who carried out research into bird migration in the 1950s, described this ability as, "...bi-co-ordinate navigation..." because the birds' use longitude to work out the sun's position and time of day, and then work out latitude from the sun's altitude. In 2007 Tim Guilford at Oxford University attached geolocators to some Welsh Manx shearwaters. He discovered that in the Autumn they flew down the West coastlines of Europe and Africa before crossing the equator and heading to Brazil then Patagonia in Argentina, before flying back via the Caribbean or the East coast of the U.S.A. and back across the North Atlantic back to Wales. Remarkably, the fastest bird flew 4800 miles (7725km) in under a week. Guilford, and other researchers of bird migration realised that, like early European navigators and explorers, the shearwaters use the wind, especially the westerlies. They not only have an innate understanding of how the wind works and where the wind will take them, but they are also able to adapt their knowledge. Portuguese researchers, Mana Dias and Jose Grandadeiro studied Cory's shearwaters on a tiny island between Madeira and the Canary Islands, called Selvagern Grande. Cory's shearwaters do not breed until they are nine and spend the preceding years exploring the oceans. Dias and Grandadeiro studied one bird, who was about five at the start of the study, over two years. It visited Patagonia, the central South Atlantic, South and North- West Africa, as well as the North-West Atlantic and the Canary current. It travelled an incredible 67,000 miles (107,826km) in two years. All the places it visited are overwintering sites for Cory's shearwater. From their research, Dias and Grandadeiro concluded that older birds chose which routes to take to their overwintering grounds, based on research they had done when they were younger. Not only did they understand the winds, but they also were able to create an internal map, using their knowledge and memory of different coastlines. However, unlike pigeons, shearwaters and their cousins, albatrosses, do not use the earth's magnetic field.

This family of seabirds live a long time, up to sixty years, and remain faithful to a mate and a breeding site. Scientists think that they might do this by using their sense of smell. Shearwaters are smelly birds, with a strong smell permeating their plumage, nest, and eggs. A researcher at the Swarthmore Institute, Julie Hagelin, has discovered that most species of seabirds have a unique smell. Seabirds in the Procellariiformes family (which means the order of the storm birds) are not the only seabirds to have an exceptional sense of smell. Gabrielle Nevitt at the University of California discovered that seabirds are able to smell dimethyl sulphide (D.M.S) which is emitted by phytoplankton. As krill, which are small crustaceans which feed on phytoplankton but are an important food source for many seabirds and other marine animals, eat the phytoplankton dimethyl sulphide is released, which the birds smell. The birds then head to that area because they know there will be lots of food. Nevitt noted that smaller procellariiformes were not attracted by smells associated with crushed krill or shrimp, although it is also part of their diet, although this smell did attract larger procellariiformes, such as albatross. She also discovered that procellariiformes burrows, and parents, smell of dimethyl sulphide, and this helps the bird associate the smell with home and helps with bonding between the parents and child. Even more astonishingly birds can recognise the smell of themselves and their relatives, but also of unrelated birds which look like them. Nevitt discovered that their hearts beat faster when they are exposed to a recognisable smell. She also carried out tests on another member of the family, petrels, to attempt to understand how sensitive their sense of smell was. Remarkably she found that they were able to detect the presence of tiny amounts of D.M.S. in the air. When their nostrils were temporary disabled, they were unable to find their own burrows, which were only three feet (91cm) away. She therefore concluded that the sense of smell was the most important sense for seabirds.

Another researcher, Anna Galgliardo, has taken this research into seabirds' sense of smell even further, and has discovered that birds remember different parts of the ocean based on what it smells of. She concluded that shearwaters, and other seabirds, can smell their way around the oceans. As Adam Nicholson says in *A Seabird's Cry,* shearwaters, and their relatives, "...are not just barometers of the sea but also their investigators and navigators."

The worrying footnote to this area of research is that scientists have also discovered that plastics in the sea also emit dimethyl sulphide which might explain why increasing numbers of seabirds are found to have ingested plastics, mistaking it for food.

Another bird which spends most of its life flying over the oceans is the kittiwake. It gets its name from its call. It breeds around the coastline of the British Isles in Spring and Summer. They eat sand eels, capelin, and herring, but also shrimps, worms, insects, and planktonic sea creatures. Researchers have shown that kittiwakes change their fishing patterns according to the tides. Unlike birds like gannets and puffins, kittiwakes are not particularly good divers, and can only dive one or two feet (thirty to sixty centimetres) under the surface of the sea. In the last half a century there have been huge declines in kittiwake populations across the Northern Hemisphere. Scientists are not sure what is causing the decline, but they seem to have problems breeding. In 2008 only 25% of kittiwake chicks fledged, whereas in 1986 every nest produced a fledgling. It is thought that a reduction in food is partly to blame because, like puffins, they are reliant on sand eels. Research has shown that seabirds need at least 33% of their prey to be available for their populations to remain stable, but if the amount of prey falls to under 33% their populations plummet. Kittiwakes are further hampered because in years where there are not much food older chicks push their younger siblings over the edge of the cliff, normally to their deaths.

One of the seabirds most people are familiar with are what are colloquially referred to as seagulls. In fact, there are many species of gull [74] but the ones normally seen at the seaside are herring gulls. Despite being

---

[74] 34 species in the world

highly adaptable, and being predators and scavengers, gull populations have declined over the last thirty years, with herring gulls declining by 50% since the early 1990s. This might partly be because in recent years ornithologists realised that there were several species of gulls which looked remarkably like herring gulls so there might have been an error counting them in the past. However, climate change and overfishing also seem to have contributed to their decline. Some species of gulls, particularly common gulls, and lesser black-back gulls, have adapted to urban living, scavenging on the rubbish left behind by humans. Almost two thirds of gulls now live in our urban connotations. Peter Rock at the University of Brighton, who has studied urban gull populations for about forty years, thinks that urban areas are attractive to gulls because they provide plenty of food, there are fewer predators (although some councils use falcons and hawks to control pigeon and gull populations), they can forage for food at night because of streetlights, and the temperature is warmer in urban areas than it is at the coast. However, research has shown that urban gulls produce less chicks. The other problem with urban gulls is that councils complain about the damage done by their guano (poo).

Attitudes to bird guano have changed over the centuries. In 1802 the Prussian explorer Alexander von Humboldt began investigating the uses of guano and his articles on the subject became extremely popular in Europe. Its potential use as a fertilizer was given a further boost by the well-known chemist, Humphry Davy, who sold a bestselling book about it entitled *Elements of Agricultural Chemistry* in 1813. He highlighted that Peruvian guano contained high levels of nitrogen, phosphate, and potassium, which help plants grow. European guano was not viewed as being of the same quality because it was affected by high levels of rainfall and humidity. The book was so popular it was translated into German, Italian and French. Bird guano was also used in the production of gunpowder. In the 1840s the Peruvian government negotiated a lucrative deal with other countries, including Great Britain. Guano was removed from three tiny islands called the Chincha Islands off the coast of Peru, where the guano was 200 feet (61m) deep. Ships were loaded with the precious cargo. The crew were careful to open windows and doors, to prevent them becoming overcome by the toxic fumes. Such was the extent of the trade that 90,000 Chinese men were brought to the islands to mine the guano with pickaxes. In its heyday three hundred ships per year visited the island to export the valuable commodity. In one year at its peak 700,000 tonnes of guano were exported from the islands. The guano was mined at such a rate that within about thirty years all the guano had been excavated. This had a detrimental effect on the topography of the island, but also to the seabird colonies who lived there. It is also believed that the guano might have been the vehicle for the introduction of a virulent strain of potato blight which decimated potato crops in Ireland in the mid-19th century, leading to the Great Famine in Ireland. Another consequence of the insatiable demand for guano was that the U.S.A. passed the Guano Islands Act in 1856 which gave American citizens a claim to unclaimed islands if they discovered guano on the islands, with exclusive rights over the mining of the deposits. The following year they began annexing uninhabited islands across the Caribbean and Pacific, one hundred in total, some of which are still American territories. Other countries, including the United Kingdom, also expanded their empires, laying claim to islands to mine the guano deposits. The beginning of the end for the trade was just before the First World War when a German chemist called Fritz Haber began producing synthetic fertilizers on a large scale.

However, that is not the end of the story. In 2020 C.N.N. reported that guano could be worth one billion dollars annually. Part of this was to highlight the plight of seabird populations worldwide, but also its use as a fertilizer because scientists have discovered that guano might play an important role in restoring marine and terrestrial habitats. In coral reef ecosystems the nutrients deposited from guano can increase reed fish biomass (a measurement of the number of fish in their reef and their size) by 48%. It will be interesting to see whether guano does become a key part of the conservation of species and the restoration of habitats.

Meanwhile in the United Kingdom, gulls are like marmite you either love them or hate them. I must admit to loving these highly intelligent, charismatic birds. Gulls are very sociable, living in colonies and helping each other. They breed with the same partner for life, up to twenty years. They use body movement and a variety of calls to communicate with one another. Gulls have two nifty tricks. One is tap dancing in order to imitate rain, encouraging worms to come to the surface. Their other talent is being able to drink seawater, something very few birds can do. They do this by using glands above their eyes which help remove excess salt from their bodies through openings on their bills.

Cormorants and shags had a bad reputation too. In medieval times black birds were always associated with the Devil and evil. However, during the 17th century French and British monarchs kept cormorants as pets because they like to watch them catch fish. Chinese fishermen also tame cormorants and use them to help them catch fish.

These are two of the least evolved seabirds, so they give us an insight as to what the original ones 100 million years ago looked like. They look remarkably similar, black with long tails with a touch of white and yellow around the bill. Shags are smaller and slimmer than cormorants. Their habitats are different, with shags preferring rocky coasts and islands, whereas cormorants also like inland freshwater waterways such as rivers and lakes. Cormorants are pretty much widespread across the British Isles. However, according to the R.S.P.B., there are more shags than cormorants. Shags prefer the Scottish, Welsh, and Irish coasts as well as the North-East and South-West of England.

Cormorants and shags can often be seen with their wings outstretched. Unlike other seabirds their feathers are not fully water resistant. Birds trap air in their feathers to keep warm but this can be a problem with birds that live in water because they are too buoyant. One solution is to be like the gannet and dive deep down under the water's surface because the water pressure squeezes the air in between the bird's feathers, making it less buoyant. Some diving birds breathe out just before they dive in order to decrease the oxygen in their lungs, whereas others have muscles in their bodies which pull back their back and feathers as they dive, making them denser. Cormorants' and shags' wings do not work very well under water, and as they cannot fish in deep water there is no water pressure to squeeze the air out of their feathers, so to avoid a buoyancy problem cormorants and shags have feathers unlike other birds, with a much more open structure, particularly at the outer edges. This allows them to absorb water, but the air is expelled as soon as the feathers become wet. Meanwhile, the inner part of the feather is denser, water resistant, and remains full of air allowing them to keep warm.

The Wonders of the Wild Places

Also look out for eider ducks at the coast. In the North- East of England, where I live, we call these sociable black and white ducks cuddy ducks after St Cuthbert, a 7th century Northern Saint who lived on a small island off Lindisfarne, off the Northumbrian coast. Lindisfarne is still home to a population of cuddy ducks. St Cuthbert was supposedly very fond of these charismatic ducks, so much so that he was instrumental in passing laws against their eggs been taken, the first recorded instance of birds' nests being protected anywhere in the world, therefore making him one of the first naturalists and conservationists.

They get their more common name from the French word *eider a duvet* with the final part referring to their down which is soft and thermally efficient. Eiderdown was first used in beds in the United Kingdom in 1689 when a diplomat called Paul Rycaut brought the idea back from Germany. Originally, they were so expensive that people left their loved ones eiderdowns in their will. It was not until Victorian era that they began to become popular. By the 1950s most people slept under a warm and cosy eiderdown. However, by the following decade when Terence Conrad popularised the duvet, eiderdowns days were numbered. This was good news for the eider duck whose population has increased.

Estuaries and saltmarshes are also good places to look for birds especially in winter when migratory birds can be seen in their thousands. Forty-five percent of the world's oystercatchers winter along the British coastline and many more live here all the year round. During the summer you are probably more likely to see oystercatchers on moorland or farmland. Over the last fifty years they have gradually moved more inland and can also be seen on freshwater. However, during the winter, these distinctive black and white birds with their orange legs and beaks, can be seen pecking in the mud. They use their strong, flattened beaks to either prise open or hammer open shellfish.

Bar-tailed godwits are another winter migrant, flying here from Siberia and the Scandinavian Arctic. Sandpipers and plovers are two large families of shorebird. The plovers have shorter legs than the sandpipers and locate prey by sight. Sandpipers locate prey by probing in the mud. They have a special sensory nerve in their beak to help them find prey under the mud.

The Wonders of the Wild Places

Of course, there is not just nature living by the sea, a lot of creatures and plants live in the sea. Three quarters of the earth's surface is covered by oceans. The seas are teeming with life. Amazingly, there is more D.N.A in a bucket of seawater than in the human body. The oceans are vital for human life. Ocean currents create rain which in turn creates the small amount of freshwater available on earth. Phytoplankton produce most of our oxygen. The seas absorb about 40% of carbon in the atmosphere. However colder seas absorb more carbon, so as climate change warms up our oceans scientists think the sea will become a less reliable at sequestering carbon. A consequence of oceans absorbing carbon is that it makes the water more acidic. This is having an impact on sea life. In the last half a century vertebrate life in the oceans, including fish, have been reduced by a third. However, there is a caveat to this statistic because scientists know more about space than they do about life in the oceans, having only explored about 5% of the deep ocean. It is believed that there might be millions of species yet to be discovered.

What is known is that there are many species of cetaceans in the sea. This comes from the Latin word *cetus* meaning a large sea creature and is a general term applied to ninety species of sea mammals including whales, dolphins, and porpoises. New species are being found all the time with the last one being found in 2021.

Several species of whales are found in British waters including minke, sperm, and humpback whales as well as fin, long-finned pilot, Northern bottlenose, and Sowerby's, and Cuvier's beaked whales, and sei whales. The Seawatch Foundation counts the whales, dolphins, and porpoises in British waters. In 2018 it recorded the highest number of species (thirteen) and the highest number of sightings (over 500), a 50% increase on the year before. Whales are divided into two groups. Baleen whales, including humpback whales, have fibrous baleen plates in their mouths, instead of teeth, which they use to filter krill, plankton, and other small algae and fish. These vertical, overlapping plates are made from keratin, like our fingernails. These whales are also known as the great whales because they are all long and heavy. They included the blue whale, which is the largest animal ever to have lived, at 108 feet (33m) long. The other group of whales, which includes sperm whales, are odontocetes, or toothed whales, because they are born with teeth. There are fourteen known baleen whales and 76 toothed whales. Most toothed whales are smaller than baleen whales, with sperm whales being the largest and heaviest, growing up to 67ft (20m) and weighing 45 tonnes. Toothed whales unsurprisingly have a bit more choice when it comes to food compared to baleen whales. They can eat everything from fish and squid to seals and sharks.

Toothed whales, together with dolphins and porpoises who also have teeth, have different sized teeth and different numbers of teeth. They even use their teeth for different purposes. For example, killer whales use their teeth for grabbing hold of prey, whereas narwhals use their tusk, which is in fact a long external tooth, as a sensory organ which may help them 'taste' the surrounding water. However, rather surprisingly, none of them actually use their teeth to chew food, preferring to swallow the food whole or in large chunks.

Toothed whales and dolphins use echolocation to look for prey and to navigate the oceans. Echolocation is the ability to use sound to 'see' the environment and it is particularly useful for species that are nocturnal or live underneath the sea where there is extraordinarily little light. In seawater sounds travels five times faster than through the air. Sounds bounce off objects providing the animal, in this case whales and dolphins, with a lot of information about the size and density of the object, where it is and whether its moving, depending on how long it takes for the sound wave to be received.

Animals which use echolocation have a transmitter (their mouth) and receivers (their ears). Their ears are positioned slightly apart which enables them to hear the sound at different times and at different levels of loudness, which helps them work out in which direction the object is.

Recent research by Ellen Coombs at the Natural History Museum in London, which involved comparing fossilised whale skulls to the skulls of living whales, has revealed that most toothed whales' skulls are asymmetrical because all their internal organs in their skull are squashed on the left-hand side of the skull, allowing soft tissue, called the melon, on the right-hand side. Wu- Jung Lee, a senior oceanographer at the University of Washington Applied Physics Laboratory, notes that the melon makes the sound clearer, by reducing the resistance to soundwaves.

Another fatty deposit which stretches from the whale's lower jaw to its ear also helps to clarify the echo. Ellen Coombs also discovered that some whales, like narwhals and the deep-diving sperm whales, rely more on echolocation so have particularly lopsided skulls.

Baleen whales do not echolocate, but they have another skill, they sing. They emit a series of long, low frequency noises, repeating phrases. These songs can last for about thirty minutes, although humpback whales can sing extraordinarily complex songs which can last for days. The songs are a way of communicating with other whales, and can be heard over long distances, but humpback whales only sing during the mating season. It is thought that their songs communicate their health, fitness, and youthfulness as well as helping to attract a female humpback whale. Humpback whales' songs can span seven octaves, almost the range of a piano. Scientists still do not know very much about whale song, but they have discovered that all the members of the same species of whale sing the same songs, which evolve over time. However, the songs do differ according to which ocean the whale resides in, so humpback whales in the Atlantic will have a different song to humpbacks in the Pacific.

Recent research by John Nabelek at Oregon State University believes that whale song could be used to help study the sediments on the rocks which make up the earth's crusts. Whilst studying earthquakes in the North- East Pacific he realised that every time fin whales sang, "...there was a response from the earth." He thinks this is because the whales' songs travel through the earth as seismic waves before being reflected and refracted by the ocean sediment and under the earth's crust. Nabelek believes that if seismometers were placed on the ocean floor they could pick up the fluctuations of the fin whales' calls, potentially providing seismologists with detailed information about the thickness of the crusts and other information as well as helping them understand faults which cause earthquakes and tsunamis. Nabelek believes that the songs of other whales might also be used for the same purpose.

Humpback whales are rarely seen in British waters but in 2013 one was seen off the Norfolk coast for the first time since records began in the 18th century and in 2018 they were seen off the North-East coast of Scotland and the Yorkshire coast. Humpbacks migrate 6214 miles (10,000km) between their feeding and birthing grounds. They spend half the time in colder waters in higher latitudes, feeding and putting on weight. They then head to warmer, shallower, tropical waters to breed.

Humpback whales are about the size of a bus with a hump in front of their small dorsal fin. They have enormous pectoral flippers which they slap on the water to attract prey, and which help regulate their body temperature. Scientists can identify individual humpbacks by looking at markings on the underside of their tails. They have bumps called tubercles on their heads. These tubercles contain a hair which acts like cat's whiskers, as a sensory tool, providing the whale with information about their surroundings. They use their baleen plates to consume 21 stone (1360 kg) of food a day.

For centuries humpback whales were hunted, and it is thought that at least 300,000 were killed during commercial whaling, which was banned in 1986, although some countries still do it under the guise of 'scientific whaling,' and since then their numbers have steadily recovered.

Most minke whales spotted in British waters are off the West coast of Scotland. There are actually two species of minke whales. The ones seen off the coast of Scotland are Northern minke whales, and they are the smallest whale seen in British waters. They are nosy animals so often come near boats. Minke whales are baleen whales, gulping large mouthfuls of fish. They have an incredibly loud song, 150 decibels, as loud as a jet plane taking off.

Minke, like other species of baleen whales, tend to swim in circles around a school of fish, decreasing the diameter of the circle until the school are in a small dense ball, then lunging at their prey. Other species of whales, such as humpbacks, blow bubbles in a circle around their prey. The prey does not dare pass over the bubbles and are therefore easier to catch.

Unlike some other species of whales, minke are quite solitary. However, they are often observed near areas where there are a lot of sea birds. Scientists think that they use the birds to help them locate food. They are nicknamed 'stinky-minkes' because they emit an unpleasant smell when they eject water.

Sperm whales have the largest brain in the animal kingdom, five times the size of a human brain. However, most of their head is filled with a yellow oil called spermaceti. Early whalers thought it was sperm, hence the name of the whale. Scientists are still not sure what this oily substance, which hardens when cold, is for but it is thought that it might help the whale to adjust its buoyancy so that it can dive deeper and rise again. Sperm whales can dive up to 3280 feet (1km) which is about the height of Scafell Pike, England's highest mountain. They can hold their breath for an astonishing two hours, although they normally hold it for 45 minutes. Sperm whales eat up to a tonne of fish and squid a day. They have 52 coned shaped teeth which weigh a kilogram (2lbs) each.

Sperm whales are sociable animals living in pods of up to twenty including juvenile whales (called calves). Males migrate to higher latitudes but head back towards the equator to breed. They have a sixteen-foot (5m) tail which they use to propel them through the water at 23mph.

Like humpbacks sperm whales were much sought after by commercial whalers for their oil and ambergris, a substance which forms around squid beaks in whales' stomachs. It is used in perfumes. The mythical albino sperm whale in *Moby Dick* was based on a real animal which whalers called Mocha Dick. The author of *Moby Dick* was Herman Melville who was part of the crew of the whaling ship *Acushnet* which sailed from Massachusetts on a voyage to the South Seas. When the ship moored at the Marquesas Islands, now part of French Polynesia, Melville and a crewmate jumped ship and were captured by cannibals before somehow escaping and registering on another whaling ship called the *Lucy Ann*. He took part in a mutiny and ended up in a Tahitian prison before escaping again, boarding another whaling ship back to America, ending up in Hawaii.

Fin whales are the second largest whale, weighing an incredible 1300 tonnes and growing up to 90 feet (27m) in length. They are solitary whales. They consume up to two tonnes of krill a day, swimming with their mouths open, then expelling the water and trapping the krill in their baleen bristles. They have asymmetrical pigmentation with one side of their jaw being white and the other black.[75] Scientists do not know why this is. Sometimes fins breed with blue whales so hybrids exist. These whales can live an exceedingly long time. The oldest one recorded was 111. Unfortunately,

---

[75] This is only found in fin whales and Omura's whales.

though, these are still hunted in some parts of the world.

Long- fined pilots got their name because each pod was thought to be guided by a single leader. Scientists have now found this is not the case, but the name has stuck. They are very sociable and are often seen swimming with other whales and dolphins. They live in multi-generational, tight-knit groups. Long- finned pilots eat squid and octopus by sucking them in.

There are two species of beaked whales in U.K. waters, and they are both named after someone. Sowerby's beaked whale is named after James Sowerby who was an English naturalist and artist who studied the skull of a whale which had stranded in the Moray Firth in North-East Scotland in 1800. Sowerby named the whale *bidens* after its two teeth, which is common in this genus. They can be found from Nantucket to Labrador in the Western Atlantic and from Madeira to Norway in the Eastern North Atlantic. They are very reclusive, rarely seen whales so scientists are not sure how many of them exist. Cuvier's beaked whale is named after a French zoologist, Georges Cuvier, who described the first fossilised skull found in the 19th century. At the time he thought the whale was extinct. They are very distinctive whales with black rings round their eyes, like a panda, and two teeth which grow as they get older, so older whales have two tusks which they use to fight over females. Consequently, they are often covered with scars, which can be visible for twenty years. The Cuvier's beaked whales' claim to fame is that it is the deepest diving mammal in the world, diving to depths of 10,000 feet (3km) and being able to be under water for just over two hours at a time. They have flipper pockets which allows them to tuck their flippers in, making their bodies very streamlined, a useful adaption which helps them dive. Like Sowerby's beaked whales, it is difficult to record their numbers because they live far out at sea and only spend short periods at the surface. However, it is known that they are affected by military sonar. If they are frightened by man-made sounds they can get 'the bends' (decompression sickness) because it makes them come to the surface too quickly.

Sei whales get their unusual name from the Norwegian word for pollack and are also known as the pollack whale or sardine whale because they are skimmers who eat at lot of these fish as well as krill, squid, and zooplankton. Northern bottlenose whales, not to be confused with bottlenose dolphins, live off the West coast of Britain and Ireland, where they eat deep-water squid and live in pods where they have long-term relationships, sometimes with a same-sex partner. During the 19th and 20th century 65,000 Northern bottlenose whales were hunted for their oil and for pet food. As they are loyal, compassionate creatures who chose to stay with wounded companions they were unfortunately easy to hunt.

Whales play an important role in the marine ecosystem by providing at least 50% of the oxygen which sustains fish stocks. They could also play a part in the bid to limit climate change because one whale can capture as much carbon as 30,000 trees can per year. Also, whale poo acts as a fertilizer for microscopic algae called phytoplankton which not only provides between 50-85% of the oxygen that we breathe (depending upon which source you read) but also absorbs millions of tonnes of carbon. Even after death the whale is a critical part of the marine ecosystem by providing food and shelter for many species on the seabed and sequestering huge amounts of carbon.

When is a whale not a whale? When it is a killer whale. Although these are in the toothed whales order they are actually the largest dolphin. Ancient sailors called them *orcas asesina ballenas* which means whale killer, and this was later flipped to killer whale. Their Latin name is *Orcinus orca* which means kingdom of the dead. Recently zoologists have realised that there are actually lots of sub-types (ecotypes) of orcas. Killer whales are apex predators of the oceans hunting everything from fish to seals to sharks and octopus to other cetaceans. However, they are picky eaters. If their family likes a particular food, they stick to catching that prey even if other prey is available. They are one of the most widely distributed species in the world, only outdone by rats and humans. They must remain conscious even when they are asleep because, unlike other mammals, they must actively decide to breathe. They do this by closing half their brain at a time and keeping one eye open.

Orcas are highly intelligent and adaptable. Not only do different groups of orcas have their own distinctive dialects, which can be as different as English is to Japanese, but they can also speak other marine species languages. They have an excellent memory and the ability to teach things to members of their family.

The Wonders of the Wild Places

Orcas are not the only dolphins who are intelligent. Lori Marino, an expert on cetacean neuroanatomy at Emery University in Atlanta, believes that dolphins are the second most intelligent creature on earth, with humans beings been the most intelligent. Marino has studied bottlenose dolphins and has noted that they have larger brains that humans and have a body-to-brain ratio greater than great apes. She also discovered that dolphins have an extraordinarily complex neocortex which is the part of the brain which is responsible for problem-solving, self-awareness, and other traits associated with human intelligence. They also have Van Economo neurons which are those associated with emotions, social cognition, and the theory of the mind, which is being able to sense what others are thinking. A recent study by Dr Blake Morton, a psychology lecturer at the University of Hull supports this. He also studied bottlenose dolphins and concluded that they had human personality traits – openness, conscientiousness, extraversion, agreeableness, and neurocism. However, he concluded that whilst dolphins had similar traits, they were not exactly the same as human traits. Bottlenose dolphins also have individual names for each other, using a unique set of whistles to attract another dolphin. They are also one of the few animals to recognise themselves in a mirror and have even been observed using tools. In Shark Bay in Western Australia bottlenose dolphins fit sponges over their beaks to protect them from sharp rocks.

Bottlenose dolphins are probably the best-known ones in the United Kingdom mainly because they are very acrobatic. When I visited Fort George on the Moray Firth in North-East Scotland, I watched bottlenose dolphins jumping out of the sea, tossing fish up into the air, before catching the fish, doing a somersault, and going back under the water. They did this for about thirty minutes and upon excitedly informing the staff at the fort about what we had seen my sister and I were told that it was a regular occurrence. I have also recently seen them off the coast at Whitby and was advised by the lifeguard that they are regularly spotted off the Yorkshire coast. Bottlenose dolphins have also been observed trying to help humans who are drowning or have been attacked by sharks. These intelligent, sociable dolphins have markings on their dorsal fins which are as unique as fingerprints, making it easier to study them. Look out for them off the coast of South-West England, the West coasts of Ireland and

Wales, the Yorkshire coast, and the Moray Firth.

Another dolphin seen off the South-West coast of England, Western coast of Ireland, and North-West of Scotland is Risso's dolphin, named after Antoine Risso, a friend of Georges Cuvier, who was the first to describe it. I saw one off the coast of Skye. At first, I thought it was a whale because of its size, but identified it by its distinctive grey scarred body. I was lucky to see it because these dolphins prefer deep water where they feed on squid.

In 1994 it was discovered that there were two versions of common dolphins – short and long beaked. They are slender with grey on top and whiter underneath and distinctive hourglass patterns on their sides. They feed close to shore so are easier to see than some other species. They live in large pods, sometimes with other marine mammals and seabirds, and, like the bottlenose dolphins, are acrobatic. Look out for them in the English Channel, and off the Western coast of England, Scotland, and Ireland.

Striped dolphins can also be seen, near the surface or jumping out of the water, on the Western coast of Britain. They can jump three times their body length. Atlantic white-sided dolphins are much shyer and rarely be seen close to shore. They work together to herd fish.

Porpoises are like dolphins but are smaller and do not have the pronounced beaks (snouts) which dolphins have. Unlike dolphins, porpoises have 60- 120 flatter spade-like teeth and flatter faces. There are seven species worldwide but only one species of porpoise lives in European waters- the harbour porpoise. They have a dark back and white undersides. Like toothed whales they have a melon, so use echolocation.

I have only seen dead porpoises washed up on beaches, but I have seen lots of seals along the North- East coast. There are two species of seals in the United Kingdom- the common seal and the grey seal. Despite their name common seals are actually less common than grey seals, but I saw a lot of them growing up because one of the best places to see them the United Kingdom is Teesmouth, which is near my hometown. They can be seen around the coast of Britain but there are more common seals around the Scottish coast, and on the eastern side of England. About 40% of the world's common seal population live in British waters. Common seals do not travel particularly far in their lifetimes, only venturing a few miles out to sea. They spend most of their lives at sea and eat whatever fish they can find. Grey seals are larger than common seals and prefer colder seas so are seen more in the North of the British Isles. Between 40-50% of the world's grey seal population live in British waters. However, there are more African elephants in the world than there are grey seals. Worldwide seals are affected by rising water levels and higher temperatures caused by climate change: as well as plastic pollution and persecution. Though, in recent years, their population has boomed in the United Kingdom. At Blakeney Point in Norfolk, one of their key breeding sites, there were 25 pups born there in 2001 and 2700 born in 2019. The same has happened at another of their key breeding sites, the Farne Islands, off the coast of Northumberland, where the population has increased by 50% in five years. However, this pattern is not replicated in Scottish populations who have seen a serious decline in the last thirty years. The Sea Mammal Research Unit is trying to find out why.

Grey seals eat a lot, consuming about 7kg (15lbs) of cod and 4kg (9lbs) of sand eels per day. They can dive to 1640 feet (500 m) Normally their heart beats similarly to that of a human, but when they dive they slow their heart down so it just beats a few times per minute. This combined with using oxygen stored in their muscles and blood helps them dive to these sorts of depth. Their nostrils are closed until they use a muscle to consciously open them to breathe. They have big eyes to help them see underwater and whiskers which are as sensitive as fingertips.

Female grey seals have a cry which sounds similar to a woman's voice. It is thought that the idea of mermaids originates from sailors seeing grey seals and mistaking them for hybridised women.

Seal milk contains 60% fat, so seal pups grow quickly and put on a lot of blubber which they use as insulation. Seal pups are not waterproof, and they have to wait until they lose the fluff at about three weeks before they can go into the sea, so they are vulnerable.

Many other creatures live in our seas. Some of them are weird and wonderful, such as the sea cucumber, which is actually an animal, related to the starfish. They live on the seabed. There are several different species of them, with some burrowing into the sand and using their tentacles to catch passing prey, and others which move about. They can make their bodies hard or squishy. They eat plants, dead animals, and even poo, and their poo provides nutrients for sea grass and other plants. They have an interesting way of scaring off predators. If a predator gets too close, they squirt out their internal organs in a sticky mess. That would scare me too! Fortunately, their organs regrow again in a few weeks. They also have another strange habit; they have no nose so breathe through their bottom. Some sea cucumbers have fish living inside them, coming in and out via their bottom and eating the insides of the sea cucumber.

Lobsters also live on the ocean floor. There are two types- clawed and spiny. Clawed lobsters have pincers and a crushed claw, whereas spiny lobsters do not have claws. I have always thought of lobsters as being red but actually they are only red when cooked. Whilst alive most of them are greenish-brown, but they can be white, yellow, or blue, or in some cases can be dual coloured. They never stop growing and can grow to three feet (1m). They normally feed on crabs and starfish but will eat other lobsters when food is scarce. Lobsters use their teeth, located in their stomachs, as gastric mills. If they lose a claw or leg, they can regrow another one. They can travel up to 225 miles (362km). Lobsters are extremely sensitive to water temperature.

Despite the name, jellyfish are not actually fish, they are related to sea anemone and coral. There are thought to be six species in U.K. waters. They have fantastic names with the ones seen in British waters being called moon, compass (seen off the southern coast of England), blue (seen off the South-West England and Welsh coast), lion's mane (seen off the North coasts of Wales and Scotland), dustbin lid (which is 35 inches/90 cm in diameter), and mauve stinger.

Jellyfish were on earth before the dinosaurs but have not evolved very much. They do not have a brain, heart, bones, or eyes. They are just a gelatinous 'umbrella' around circles of tentacles which they use to paralyse their prey before eating it. Their sting is coiled and is fired like a harpoon. Be careful because all the jellyfish found around British coasts have a sting and can sting you even if they are dead. Contrary to popular belief weeing on a jellyfish sting makes it worse. If you are stung by a jellyfish rinse the area with seawater or vinegar and remove the spines with tweezers or a debit card. Soak the area in very warm water (or wrap in a very hot towel) for thirty minutes and take painkillers. If the symptoms do not improve seek medical attention.

They eat a variety of prey including fish, shrimp, zooplankton, crabs, and even other jellyfish. The jellyfish's mouth is central and is multifunctional. They eat and poo through the same orifice and squirt a jet of water out of their mouth to propel them forwards. They are not good swimmers and tend to drift on the ocean currents. Jellyfish are not sensitive to water temperature and are equally at home in shallow and deep water so live in all the world's oceans.

Despite being primitive they do have one interesting talent. They are bioluminescent, which means they can produce their own light. They are not unique in this ability. Many marine species, especially those living in the deep oceans, can luminesce. Kevin Raskoff, a scientist at California State University's Monterey Bay Aquarium Research Institute, estimates that about 90% of deep-sea dwellers are bioluminescent. They use it as a way of attracting mates and prey, distracting predators, and as camouflage.

The way animals bio luminate differs according to the animal. In the case of jellyfish, they need a molecule called luciferin and one of the following enzymes: - luciferase or photoprotein. When these react with oxygen, they produce energy. If an ion, such as calcium, is also present this reaction also causes light to be emitted. Some animals naturally contain luciferin whereas others need to acquire it from the food they eat or, in the case of species such as the bobtail squid, by having a symbiotic relationship with bioluminescent bacteria. In the case of jellyfish, the luciferin comes from the small bioluminescent crustaceans they digest.

Many species of fish are also bioluminescent. Risso's lanternfish (named after the same person that Risso's dolphin is named after) which is found in oceans all over the globe, except the poles, and can be found off the South and West coasts of Britain, is a small fish which normally swims in shoals. The fish has light - producing cells, called photophores, under its eyes and along its lateral line. It is thought that it uses bioluminescence to communicate and to ward off predators. Kitefin sharks and velvet-bellied lanternsharks, both of which can be found in U.K. waters, are also bioluminescent. Sharks use a different technique to produce light. The velvet-bellied lanternshark has thousands of photophores on the lower part of its body and its dorsal fins. These are made from clusters of light-emitting cells called photocytes. Dr Julien Claes and Dr Jerome Mallefet and their team at the Catholic University of Louvain in Belgium have studied how the velvet-bellied lanternshark controls its bioluminescence. They have discovered that they control the photophores by hormones and neurotransmitters. The hormones they use are melatonin and prolactin. The former produces a long-lasting glow which can last for several hours. The latter produces a peak brightness after twenty minutes and only lasts for about an hour. It uses melatonin to help it to 'disappear.' It does this by counter-illumination. The shark produces light similar to sunlight, which removes the shark's silhouette from anything looking up at it. When the lanternshark wants to communicate, attract a mate, confuse prey and predators it activates the prolactin hormone, so it has a short, bright light along its dorsal fins.

People might be surprised to learn that there are forty species of sharks in U.K. waters, with about eleven living here all the year round. The largest of these is the blue shark which is about thirteen foot (4m) long, and lives in the sea off the South- West coast of England between June and October. They are migratory species. In the Atlantic they migrate in a clockwise direction, following the gulf stream from the United Kingdom to the Caribbean and returning by following the Atlantic North Equatorial Current. They travel 5700 miles (9173km) a year in single sex groups, swimming at 30 mph. Unlike bony fish they do not have a swim bladder to keep them afloat so have to constantly swim. They are apex predators, eating tuna and swordfish, and occasionally whales and other sharks, although they are sometimes eaten by orcas, crocodiles, and other sharks. During mating in the autumn, the male blue shark bites the female, so her skin is three times thicker than the male's.

Between May to October, it is also possible to see Britain's largest fish, the basking shark, off British coasts. I saw them as I walked along the coast in Cornwall. As well as Cornwall they can also be seen off the coast of the Isle of Man and the Inner Hebrides. They are easy to spot because they are huge. They weigh the same as a double-decker bus. However, do not worry, they only eat plankton. They are passive feeders. The do not actively hunt but just swim with their huge mouths open, taking in water, which is then pushed out through their gills with them eating whatever is left. Sadly, their numbers are declining due to hunting.

Sharks have been on earth longer than trees, 420 million years. They are fish but, unlike fish, they have cartilaginous skeletons, rather than bones, which helps them move quickly through the water. Unlike fish, they do not have scales and are unable to swim backwards. They have stiff fins which are not controlled by their muscles. They stay buoyant because of their cartilaginous skeletons and a really oily liver. The latter helps them stay balanced in deep water. Sharks have incredible senses. They have a fantastic sense of smell and have such good hearing that they can hear prey 3000 feet (914m) away. Two thirds of their brain is devoted to smell. Their eyes are on their side so they can have a wider view of their surroundings, but scientists believe that they are colour-blind. However, sharks that live in cold waters have the amazing ability of being able to heat up their eyes, using a special organ behind their eye sockets, in order that they can hunt more efficiently. Sharks who live in deep water have blue eyes to help them attract as much light as possible, whereas sharks that live nearer to the surface of the sea have darker eyes, to protect them from the sun. The average shark has between 40-45 teeth in up to seven rows. They regularly lose their teeth and can have 30,000 teeth during their lifetime.

They use electroreception to navigate and locate prey. This is a sense which many aquatic cartilaginous animals, and other animals such as some birds, have had for hundreds of millions of years. The animal can detect weak electric fields by using either specialised receptor cells or, as in the case of pigeons, interacting with magnetite crystals in the earth's geomagnetic field. Animals distort electrical fields as they swim through the seawater and produce electrical currents every time they move. The electricity travels through the seawater through sodium or chlorine ions. Fish have a different charge to the seawater solution in which they swim, this creates a weak voltage, like a battery. Sharks can sense tiny changes to the electric current, as little as one-billionth of a volt. It uses this ability to detect and catch prey. The shark detects the changes in the electrical currents through tiny dots around its mouth. These are known as the ampullae de Lorenzini. They are filled with an electrically conductive jelly. The ampullae are lined with hairlike cells(cilia). When the cilia detect changes in the electrical currents, they trigger neurotransmitters in the shark's brain which informs the shark that a living animal is nearby. When they are about three feet (one metre) from their intended prey, electroreception causes them to orientate their jaw, for an accurate attack. As they near the prey they roll their eyes into their heads, to protect them, and are just guided by electroreception. It is thought that this also explains why some species of sharks keep attacking a human victim, because the salt from their blood intensifies the electrical field around the victim. Do not worry. Out of the more than 500 species of sharks in the world only 34 species will attack humans. Shark attacks are incredibly rare- less than a hundred people per year, and of these only 10% are normally fatal.

The electrical current in the seawater also interacts with the earth's electromagnetic field, producing electrical fields which the shark, and other aquatic animals, can use to help them navigate.

Skates and rays are closely related to sharks. They also have cartilaginous skeletons and no swim bladder. The common skate is the largest species of skate in the world, with a nine-foot (3m) diameter. Unfortunately, despite the name, they are now critically endangered because of overfishing. They can be found in the United Kingdom in deep sea lochs in the West of Scotland and West of Ireland.

In the 16th century dried out common skates were used to make fake sea monsters because their mouths and nostrils could be used to form a humanoid face. These grotesque curiosities were originally called Jeune D'Anvers by French sailors, but this was anglicised to Jenny Haniver. Many people believed that the Jenny Hanviers were real monsters, but the father of modern zoology, a Swiss biologist and natural historian, Conrad Gessner, was wise to it and advised people in 1588 that they were just preserved skates. However, he did not get everything right. He strongly believed that unicorns existed.

The common stingray is normally found in (sub) tropical seas, but it can also be found in the English Channel and Irish sea. They like shallow water, where they live on crabs, worms, and shellfish as well as scavenging on dead fish. They have a venomous stinger in their tail. It can be up to fifteen inches (38cm). It is made of a cartilage material called vasodentin. The underside of the stinger has cells containing a protein-based venom.

Once the part of the stinger pierces the flesh the cells break releasing venom into the unfortunate victim. The stinger normally snaps off when they attack, but fortunately they can regrow it. Generally, they will only use their venomous stinger as a last resort. They can kill a human. The Australian naturalist Steve Irwin is probably the most notable victim of a stingray. However, people normally recover if they seek immediate medical attention.

Electric rays can be found in the English Channel, Celtic Sea; Welsh, and Northern Irish coasts, and Western Scottish coasts, but they are difficult to spot because they live in deep water and are nocturnal. There are twenty ray species which give an electric shock, but this one gives the largest and most powerful shock. They generate electricity through muscle contractions and store electricity in two kidney-shaped organs. The electricity is sent out of their body. A large electric ray (they can grow to six feet/ two metres in diameter) can give a shock of 220 volts. This apparently is not enough to kill a healthy human but is enough to kill a horse. It uses its ability to give an electric shock both as a defensive mechanism and as a way of catching prey. If they can get close enough to prey, they wrap themselves around their victim and give it multiple shocks. Scientists believe it has two shocks – a low-voltage one which acts as a warning, and a more powerful one to catch prey or ward off predators. It is thought that it has these two options because a powerful shock uses up all its electricity and it then must allow time to regenerate, making it vulnerable to predators, whereas the low-voltage shock allows it still to have plenty of electricity in reserve.

The electric ray's Latin name is Torpedo nobiliana. It was a name given to them by the French biologist Charles Lucien Bonaparte in the 19th century, but they had been known as torpedoes for centuries, thereby predate of the word's usage meaning a weapon. American inventor, Robert Fulton, is credited with inventing the first functioning torpedo as a naval weapon. He used the word to describe an explosive charge which was towed behind an early submarine which he had invented.

Bony saltwater fish can swim in schools of millions of fish. It is the fish in the middle of the school who controls it. The fish use their eyes and lateral line to hold their place in the shoal. They have special hairs along their lateral line which helps them sense a change in water pressure, alerting them of other fish and predators.

Saltwater fish need to drink more water than freshwater fish, so water is constantly flowing out of them in order that they do not dry out. They can die if there is not enough oxygen. Fish do not have vocal cords so communicate by making a variety of noises using different parts of their body, such as rattling their bones and gnashing their teeth. Most fish can see in colour, and some can also see in ultraviolet light.

Fish come in a variety of colours which they produce through pigments in cells called chromatophores. They have three colour pigments- erythrin (red), melanin(black) and xanthin(yellow) which occur in different chromatophores. The pigments are obtained from eating shrimps, algae and snails and other aquatic prey.

If the fish has no chromatophores it will either look white or appear to be the same colour as its background. The distribution of the pigment is affected by the nervous system and hormones, background colour, water temperature (cold-water fish are at their brightest in the autumn when the cold temperature causes the pigment to spread throughout the cell), and water quality. High levels of ammonia, nitrogen, and chlorine in the water can affect the pigment, making the fish appear paler or darker depending in the levels of concentration in the water. Fish also have iridocytes on the surface of their skin which reflects the light, giving scales a silvery appearance.

There are estimated to be about one hundred saltwater fish species in British waters. They are broken down into seven main categories: round fish (which are a classic fish shape, such as cod), flatfish, skates and rays, sharks, mini-species, off-shore and deeper water fish, deep-sea fish, and then a broad category encompassing rare and unusual fish.

A sunfish falls into the last category. They are normally found in tropical seas around the equator, but because of warming seas they can now also be found off the South and West coast of Great Britain. They are the heaviest bony fish in the world, weighing just over two tonnes (2000kg) and measuring 14ft by 10ft (4m by 3m). They mainly eat jellyfish which they suck in with a strange beak which they are unable to fully close. Their teeth are in their throat, so they can break food down before it reaches their stomach. One female sunfish can lay up to 300 million eggs a year.

Atlantic bluefin tuna is another fish which is now seen in British waters (especially off Southern and Western coasts) due to warmer seas. Incidentally, the increase in the temperature of the seas might not necessary all be caused by climate change. It could be because of the Atlantic Multidecadal Oscillation which is a naturally occurring long-term change to the temperature of the North Atlantic which occurs every 60-100 years and oscillates between a negative phase, when the seas get cooler, and positive phase, when the sea warms up. These migratory fish can swim at 40 mph and use their speed to catch mackerel, herring, and sardines. In the United Kingdom the tuna you eat will probably be skipjack tuna which is so common it is known as the 'rat of the sea.' However, in Japan bluefin tuna is consumed. In 2019 a sushi restaurant owner called Mr Kiyoshi Kimura paid an incredible £2.5m for a 44 stone (278kg) bluefin tuna.

The grey triggerfish is found of Southern coasts of England in the summer and is moving northwards. It uses its powerful jaws to crunch the shells of crustaceans. It has spiny dorsal fins. The first fin is raised and the second fits into a groove to keep the spine up. The second part can be pressed like a trigger, hence the name, in order to lower the spine.

Another unusual fish is a lumpsucker which has a sucker which it uses to stick itself to rocks. It is a deep-water fish scavenging on food on the seabed. It is particularly found off Northern coasts of Britain.

Another very primeval deep -sea scavenger living in British waters is the hagfish. It is closely related to lampreys. It lives at depths of 4921 feet (1500m) and has very primitive eyes which lack lenses and cannot focus but can differentiate between light and dark.

They have lived on the seabed using their disc-like jawless mouth to bore into dead and rotting fish for at least 500 million years. They also live inside sharks and whales, absorbing that animal's nutrients through their skin. They can last for many months without food.

Another name for the hagfish is the slime eel although it is not an eel. It is difficult to clarify because it does not have the hinged jaw of a fish or the cartilaginous skeleton of a shark or ray. They have a hundred slime glands on the side of their body which produces a milky substance which is a combination of mucus and thread, which when it comes into contact with water becomes slime in a matter of seconds. It is thought that they use it as a defence mechanism against sharks because the slime can clog up the shark's gills, suffocating it. Each hagfish has hundreds of miles of slime thread inside them.

Tim Winegard at the University of Guelph in Canada is studying the fibres of hagfish slime. The slime is very thin but also extraordinarily strong and stretchy. When it is dry it looks like silk. Tim Winegard and his team believe that hagfish slime could be a natural and renewable alternative to synthetic fibres such as nylon and lycra, and hope that it might be used to make breathable sportswear and even bullet-proof vests. They are also investigating whether it could be used as an alternative to plastic. At the current time they are trying to work out how to do it. There will not be hagfish farms any time soon because hagfish do not breed in captivity. Winegard and his team are looking at reproducing the proteins found in hagfish slime artificially in the laboratory. The same thing has been attempted with spider silk, but it was difficult to reproduce, but Douglas Fudge of the University of Guelph thinks that it might be easier with hagfish slime because the protein is smaller, so should be easier to replicate. Other members of the team are trying to make threads out of genetically engineered bacteria.

If you are now suitably impressed by this primitive fish maybe you would like to celebrate Hagfish Day which is the third Wednesday in October. Alternatively, you could vote for it to be our national fish. In the 1980s Norway held a poll to find its national fish, expecting cod, herring, or haddock to be voted the winner. However mischievous students, thinking it would be funny to disrupt the poll, voted in their thousands for hagfish, making it the winner. The people running the poll did not see the funny side so unfortunately the result was declared void.

A more familiar fish is a halibut whose name means holy flatfish because it was only eaten on holy days. They are the largest flatfish, growing up to a massive fifteen feet (4.5m) long. They can be found in the North Sea but are sadly endangered because of overfishing. If you would like to find out more about them visit the Kveitmuseet on the Norwegian island of Senja which is a museum dedicated to the humble halibut.

The Wonders of the Wild Places

I do not think there is a monkfish museum but maybe there should be. It lies camouflaged on the seabed. It has a protuberance called an esca on its head which it uses to attract prey. If an unfortunate fish or crustacean does come over it snaps forward, quickly consuming the prey in its mouth. Its teeth are hinged so the prey is held securely in its jaws, with no hope of escape. It has an extendable stomach so can eat prey, such as octopus and squid, which are twice as big as it.

Cod and haddock are popular fish to eat battered with chips, but it could be argued that it is the herring which has the most cultural importance in the United Kingdom, especially along the Eastern coasts of Britain. These are an extremely abundant fish eaten by humans for at least 5000 years, and many other sea species too. They tend to swim in shoals, some as big as one cubic mile, containing billions of fish. They migrate along the Eastern coast of Britain. Until the 1960s groups of women, known as herring girls, would follow the herring shoals, moving along the East coast over several months of a year. Herring have been fished since at least Viking times, but the Scottish herring industry declined in Tudor times because fishermen were ordered to supply residents with herring at a fixed price and were unable to export them. This led to many Scottish herring fishermen emigrating to the Netherlands. Charles II set up a Council of Royal Fishery in 1677. The Council compelled all coffee houses to buy a barrel of herring per year. However, this did not boost the home market as Charles had hoped. Things changed significantly for the herring industry in Victorian times, when the coming of the railways made it easier for herring to be brought to market and for groups of girls, the youngest in their teens, many from the highlands and islands of Scotland, to be moved along the East coast from Stornoway in the Outer Hebrides to Great Yarmouth in Norfolk, gutting, and salting herring. They would be in Scotland and the North-East of England during the Spring and ended up in East Anglia by the Autumn, before heading back home for the Winter. At its heyday in the 19th century, 30,000 vessels were fishing for herring along the Eastern coast of Britain. The government paid £3 per ton to the owners of fishing boats which held sixty tons of herring, so it was a lucrative business. However, it was not so lucrative for the herring girls. A team of three or four girls was assigned to a particular trawler for the whole season. They stayed on dry land with two or three girls gutting the fish and one packing the fish into barrel. They got paid by the barrel so had to work fast. Before refrigeration, the herring was either smoked or salted to preserve it. The herring girls regularly got nicks off the sharp gutting knives they used, which combined with the brine and salt must have been very painful. They worked for about fifteen hours per day, six days a week.

People might not think of a seahorse as a fish, but they have gills, fins, and a swim bladder so they are counted as fish. Seahorses and razor fish are the only fish to swim upright. Seahorses are not very good at swimming so often die of exhaustion. They propel themselves by using a sail fin on their back which they must flutter up to seventy times a second. They use small fins near the back of their head to steer. They get their name from the Greek word *hippocampus* which means horse sea monster. Incidentally part of the brain is also called hippocampus because it is the shape of a seahorse.

There are 54 species of seahorses in the world, with new species still being found, mainly living in temperate and tropical waters, but we have two species living off the British coast- the spiny seahorse and the short-snouted seahorse. They are mainly off the West and South coasts of Britain but in recent years have also been found in the North Sea and the English Channel.

Seahorses have no teeth or stomach, so the food moves quickly through their digestive system, so they must eat constantly to stay alive. They consume at least 3000 shrimps per day. They catch prey by anchoring themselves to sea grass or coral with their tails and using their snouts to suck up plankton or small crustaceans. Their snouts are like a powerful vacuum cleaner. Seahorses can move their eyes independently so they can look backwards and forwards at the same time. This helps them keep an eye out for predators although fortunately not many things eat seahorses because they are very bony and indigestible.

They are the only animal species in the world where the male gives birth. This allows the female to make more eggs straightaway and reproduce more quickly. After an eight-hour courtship during which they spin around, swim side by side, and change colour, the female lays eggs into the male's pouch and he carries them around until they hatch into miniature seahorses. He will give birth to up to 1500 baby seahorses but fewer than 0.5% of them will make it to adulthood.

Seahorses can change colour and mimic aquatic plants by using a fleshy appendage called a cirri. They change colour by using neurons and hormones to switch chromatophores on. They are not the only species which can do this. Cuttlefish and their relatives, octopuses, and squid can also change colour to suit their surroundings. Roger Hanlon, a senior scientist at the Marine Laboratory in Massachusetts, in collaboration with scientists at the University of Sydney and the University of Melbourne, have studied camouflage strategies in cuttlefish. Their skin is like a high-definition colour television. It can combine basic colours to form complex hues and patterns. They have up to ten million small colour cells all over their skin, each controlled by a neuron. They use a pigmented organs, elastic sacs called chromatophores, to display red, yellow, black, and brown. Each chromatophore has muscles which radiate from it like spokes on a wheel. The cuttlefish contracts or relaxes these muscles in order to expose or conceal different colours. They have 200 chromatophores per 0.001 inch (1mm^2) of skin. They also have a layer of cells called leucophores which reflect white light. In shallow water it looks white but as it goes deeper it gets more green or blue, so the cells reflect those colours.

As if this was not impressive enough, they can also change the sculpture of their skin by contracting circular muscles, pushing up a liquid into little nodes, spikes, or flat blades which they alter depending on whether they want to mimic a rock or kelp. Mark Norman, the senior creator of molluscs at the Museum of Victoria in Australia explains that, "…by adding he structural component it gets rid of its profile and outline, and predators looking for that shape will become confused." Apparently the American and Australian military are looking at using a similar technique in soldier's uniforms, so they are better camouflaged in a warzone.

It is thought that the ability to change colour is innate because young cuttlefish already display up to thirteen body patterns. What makes this even more amazing is that cephalopods (a family including cuttlefish, squid, and octopus) are colour blind. However, theoretically at least, (although at the moment their research is based on a computer model of an octopus's eye so they need to test their research on a live cephalopod), Professor Justin Marshall and Dr Wensung Chung at Queensland Brain Institute have discovered the cephalopods can adapt their vision depending on the colour and depth of the water they live in. It is thought that they do this by rapidly focusing their eyes at different depths, creating a 'chromatic blur'. Each colour has a different wavelength. The cephalopod's lenses can bend some wavelengths more than others so it is possible that one colour can be in focus whilst another is still blurred. It is believed that a quick sweep of focus enables the viewer to work out the colour of an object based on when it blurs. Thus, the cephalopod can spectrally tune their visual focus depending on whether they are living in coastal waters (green) or deep sea (blue). Many cephalopods have off-centre pupils (and cuttlefish have w-shaped pupils) make this blurring effect more extreme. Scientists at the Max Planck Institute for Brain Research and Frankfurt Institute for Advanced Studies liken chromatophores to biological colour 'pixels.' When camouflaging they do not try and match their local environment pixel by pixel but extract, through their vision, a statistical approximation of their environment and use this to select a camouflage out of their large repertoire which is selected by evolution. They have further discovered that the physical arrangement of the chromatophores is irregular and is as unique as a fingerprint.

Octopus are also able to change colour, pattern, and texture. They have large human-like eyes which send signals to their chromatophores which makes them expand or contract. Scientists at the University of California Santa Barbara studied the Californian two-spot octopus and discovered that the octopus can sense light without using its central nervous system. Octopus do this by using opsins (light sensitive proteins). It is the first time this has been discovered if cephalopods. The scientists have named the process Light-Activated Chromatophore Expansion (L.A.C.E.). Desmond Ramirez, who was one of the doctoral students involved in the project, says that "...the octopus does not sense the light in the same amount of detail as it does when it uses their eyes and brain, but it can sense an increase or change in the light. Its skin is not detecting contrast...but rather brightness." Ramirez and Todd Oakley found that the octopus had rhodopsin, which is normally produced in its eyes, in the sensory neurons on the skin tissue's surface.

As if being shapeshifting, camouflage experts is not enough

cephalopods are extremely intelligent, probably the most intelligent invertebrates. Octopus and cuttlefish have huge brains, with octopus having the largest brain-to-body ratio of any invertebrate. Octopus actually have nine brains, a central doughnut-shaped brain in their heads and mini-brains, called ganglia, on each of their arms. This enables them to have localised and centralised control of their actions. In recent experiments conducted by Alexandra Schnell and her team at Cambridge University, it has been shown that cuttlefish can delay gratification, by waiting for their favoured prey instead of just eating another source of prey which is readily available to them. Prior to these experiments it was just thought that primates, parrots, and corvids, as well as humans, were able to do this. It is not known yet why they are able to exercise self-control and delayed gratification, but Alexandra Schnell thinks that it might be a by-product of their ability to camouflage themselves and wait for prey, "... so the cuttlefish can optimise foraging by waiting for better quality food". Octopus have been observed performing problem solving tasks, such as unscrewing the lid from a jar in order to obtain the food within. They have also been shown to be capable of differentiating between shapes, and observing, and learning and forming long-term memories. They are also one of the few animals to use tools and have been shown to recognise individuals, including human faces. At one aquarium one unfortunate employee was regularly squirted at by an octopus who had taken a dislike to the employee.

In 2010 Paul the Octopus became a media sensation when he correctly predicted the result of seven Germany games in the F.I.F.A. World Cup in South Africa. He made his predictions by choosing a mussel from one of two tanks, each with a different flag on it. He correctly predicted that Spain would beat Germany in the semi-final. It is not known how Paul managed to correctly predict the results and it is thought that it was probably just a co-incidence. Sadly, Paul could not capitalise on his fame because he died of natural causes shortly after the World Cup.

The common octopus, common (European) squid, and bobtail squid live in British seas. Squid use their long tentacles to catch prey. Common octopus are nocturnal and are especially common off the Southern and Western coasts of Britain. Their genera, Cephalopods, means *head-foot* in Greek. Common octopus have eight legs, each with a double row of suckers, attached to its head. It has 240 suckers per limb. They can move the suckers independently and use them to touch and manipulate objects but also to smell. Octopus taste by touching something with their suckers. If they lose a leg, they can regrow one. They have three hearts and, like the aristocracy, blue blood. Two of the hearts pump blood to its gills and the third one pumps blood through its body. Its blood is blue because it contains a copper-based protein called hemocyanin. Despite having three of them, the octopus's hearts stop beating when it swims, so it normally crawls across the seabed rather than swims. The common octopus has glands containing venom which it uses to incapacitate crabs, which are its favourite prey. The venom is generated from bacteria living inside the octopus. All the 300 species of octopus have venom but, in most cases, not enough to poison a human. However, the blue-ringed octopus, which is not found in British seas, has such powerful venom that if attacked by one, a human will become paralysed within a few minutes. Common octopus remove the crab's shell by using a toothed cartilage-based protein, which they use to drill into the crab's shell. They have soft bag-like bodies which they can use to squish into small spaces if they need to hide from predators.

Cephalopods can also emit ink to distract predators by obscuring the predator's vision and affecting their sense of smell. The mucus -rich ink can also interfere with fish's gills. Scientists studying squid ink have realised that it contains a chemical called dopamine which is a neurotransmitter which produces a sense of euphoria in human brains. They are now trying to discover how this helps the squid. The ink also contains the same chemical which makes birds have dark feathers, melanin, which dissipates ultra-violet radiation; as well as free ammino acids and metals. Cuttlefish ink has been found to be antimicrobial, anti-retroviral and it is even being examined to see if it could be used to treat cancer.

In cephalopods the ink is generated, stored, and evacuated from a specialised structure called the ink sac (which includes the ink gland) which feeds into the rectum. They eject ink from their anus and a siphon (which is a funnel shaped organ) in order to create a cloud of ink.

A female octopus lays between 200,000 and 400,000 eggs which she watches over maternally until they hatch into larvae, forming a cloud of plankton. Unfortunately, that is the last she sees of her offspring because her body performs cellular-suicide, destroying her tissue and organ, and eventually causing her death.

Why is the sea important?

The phytoplankton in the sea emits more oxygen than all the forests and jungles combined. They provide about 65%[76] of the oxygen that we breathe. We need a healthy ocean in order that humans can survive.

Sir David Attenborough has described the oceans as, "...the greatest ally against climate change." The sea absorbs 30% of excess carbon and 90% of the heat from global warming.

The ocean currents regulate our weather and our climate. Every drop of seawater rides the currents. It takes a thousand years to complete the circuit. The oceans soak up heat and transport warm water from the Equator to the Poles and the Tropics. Without these ocean currents the weather would be too extreme and less of the world would be hospitable. It regulates the earth's temperature. Ninety-seven per cent of the water on earth comes from the seas. Almost all the rain is generated by the seas.

The sea is home to 80% of all life on earth. This is the marine life that we currently know about. Biologists think that there might be as many as 300,000 marine species living in the deep which have not been discovered yet. Recently, twelve new species were discovered in the Atlantic. Sadly, these new sea mosses, molluscs and corals are already under threat from climate change. Only 80% of the oceans have been explored and all the deepest parts of the oceans have just been surveyed in 2021. As Professor George Wolff, an ocean chemist at Liverpool University, says, "We...have better maps of the surface of the Moon and Mars than the seabed. So, whenever you go to the deep ocean, you find something new – not just individual species but entire ecosystems."

---

76 There is a variation depending on which source you look at, but most experts seem to think that phytoplankton provides us with between 50-80% of the air that we breathe.

One billion people rely on the sea as their main source of protein. As the world's population is growing by 1.5 million people per week, it is likely to be even more important source of food in the future, especially as people look for alternatives to eating red meat. Sixty percent of the world's population lives near the sea and almost 60 million people work on or near the sea. Ninety percent of world trade is made by sea.

The sea also has therapeutic properties. Being near or in the sea has been shown to be good for a human's physical and mental health. I am asthmatic and I always seem to be able to breathe better when I am at the seaside. It has been shown that humans share the mammalian diving reflex which animals like dolphins and seals have, when their faces enter the water their heart rate slows and the blood flows to the vital organs. We have also turned to marine life to make vital drugs, such as the anti-viral drug Zovirax, made from nucleosides isolated from sponges, and Yondelis which is the first anti- cancer drug of marine origin.

What are the threats involving the sea?

POLLUTION

Plastics are a range of (semi) synthetic materials which use polymers as a main ingredient. Polymers are long molecules built around chains of carbon atoms, with either hydrogen, oxygen, sulphur, or nitrogen filling the spaces. Plastics were first invented by Alexander Parkes who made a material called parkesine from organic cellulose. Once heated it could be moulded and retained its shape when it cooled down. He proudly demonstrated his new invention at the Great Exhibition in London in 1862. In 1907 a Belgian-American chemist called Leo Baekeland created the first synthetic mass-produced plastic. You might have it in your home if you live in a pre-war house. It was called Bakelite. Plastic was seen as a wonderful new product which could be moulded into a variety of things and had numerous uses. Incidentally, the term plastics was first used by Leo Baekeland after the Greek word *plastikos*, which means to fit for moulding. Despite them being invented about 160 years ago, plastics have only been made, on an industrial scale, for about seventy years, but the affect that have had on the environment and the life of animals, birds, fish, and humans has been catastrophic.

The Wonders of the Wild Places

More than eight billion tonnes of plastic have been made and every piece of it still exists. Less than 10% of it being recycled worldwide. Approximately 260 million tonnes of plastic are made each year (the equivalent of the weight of the entire human race combined) and about 40% of this is single life plastic. World plastic production has doubled in fifty years and is expected to increase by a third in the next five years. In 1974 the average consumption of plastic per capita was 4lbs (2kg), now its nearly 7 stone (43kg). The covid pandemic has not helped. It led to a huge demand in plastic personal protective equipment, none of which is recyclable.

Every year about eight million tonnes[77] of plastic ends up in our oceans. Collectively human beings must take responsibility for this, but China is the worst offender, dumping one million tonnes of plastic into the sea each year. A lorryload of rubbish is dumped into the sea every minute. Eighty percent of marine pollution[78] is from land-based sources. However, there are some regional variations and for areas like the North Sea this is about 50% with the remainder coming from another big source of marine pollution, abandoned and lost fishing equipment. The European Environment Agency estimates that 640 tonnes of fishing gear, including nets, end up in the seas each year. It is thought that by 2050 there could be more plastic in the sea than fish.

The problem is that plastic does not degrade. However, a combination of sunlight, seawater, and the action of the waves breaks it down into microplastics, some of them are 150 times smaller than the diameter of a human hair. This can take a long time. For example, it takes about fifty years for a plastic cup to break down, 500 years for a plastic bottle or disposable nappy to be broken down into microplastics, and 600 years for fishing equipment to break down. However, Dr Katsuhiko Saido at Tokyo University has recently carried out studies on polystyrene, a commonly used plastic, and was alarmed to discover that at low temperatures (under

---

77 The University of Georgia, USA, thinks that the true figure is between 4.8 million and 12.7 million based on their survey of rubbish entering the sea just in 2010.
78 About 60-90% of this is plastic but it also includes oil, chemicals and fertilizers, which have ended up in rivers and then flowed into the sea, as well as glass, metals, rubbers and textiles.

30°C) it can degrade in less than a year in seawater, releasing toxins into the sea and becoming microplastics. In addition, some microplastics are in a variety of products, including toothpaste and cosmetics. The microplastics move around the ocean's currents, creating large patches called gyres. The largest gyre is in the North Pacific and is said to be twice the size of the United States of America and contain 3.5 million tonnes of rubbish, mainly plastic. There are five more major gyres in the oceans, including the Atlantic.

A story about plastic ducks, twenty containers of plastic ducks to be precise, illustrates how far plastic can travel around the world's oceans. In 1992 the ducks were lost overboard on a ship travelling from China to the U.S.A. By 1994 some of the plastic ducks were observed in Alaska and by 2000 they had reached Iceland. They have now been sighted in the Pacific, Atlantic, and Arctic oceans.

Dr Richard Thompson at the University of Plymouth describes the microplastic as 'mermaid's tears.' He estimates that there are about 100,000 of them in every kilometre square (0.6 square miles) of seabed and 300,000 in every kilometre square of the sea. The microplastics are ingested by fish, birds, and animals, and when we eat a fish or shellfish, we ingest them too. Not only does the plastic ingested by animals and birds prevent from them from feeding, ultimately causing them to starve to death, but they are also poisoned by the toxins in the plastic. Microplastics are literally everywhere. A study by an independent marine research institute called Algalita in 2004 found that there were six times as much plastic in the water samples they took than there was plankton. However the majority of the plastic (70% of it) ends up on the seabed. The Marianas Trench is the deepest trench in the World. At 36,201 feet (11,034m) it is larger than Everest. In 2014, scientists exploring the trench were delighted to find a new species of crustacean. However, they were horrified to discover that even before this new species was known to science it had already ingested plastic. They named it *Eurythenes plasticus*. Ninety percent of seabirds are also thought to have ingested plastic as well as more than 270 other species. Approximately 100,000 marine mammals per year are killed by plastics.

Plastic in the sea is harmful in other ways too. Larger marine animals, such as whales and dolphins, can become entangled in it, meaning that they suffocate because they are unable to come up for air. Seventy percent of marine animal entanglement is in discarded fishing nets. It also affects the reproduction system in fish, birds, and animals. Plastic acts like a sponge, absorbing dangerous chemicals. Scientists have also discovered that invasive species can use the plastic, like a raft, to make it easier for them to travel to other areas. This affects ecosystems.

The problem with plastic does not just affect our seas, a lot of it is washed up onto our beaches too. Dr Jenna Jambeck from the University of Georgia in the U.S.A. states that this is equal to, "...five bags full of plastic for every foot of coastline in the world."

One hundred and thirty-six tonnes of microplastics per year are ejected from the sea. In 2008 the annual International Coastal Clean Up noted that the most common items found on beaches were cigarette butts, plastic bags, plastic wrappers, and food containers. In the United Kingdom Dr Thompson has discovered that microplastics are the second most common plastic litter washed up onto our beaches. Some islands in the Pacific look as if they have multicoloured sand because there are so many pieces of microplastics on the beach. I have seen themselves when walking on beaches looking for shells and sea glass.

Microplastics are not just harmful to wildlife they affect human health too. They are in the air we breathe and the water we drink. A type of plastic called Bisphenol A has been shown to cause, amongst other things, miscarriages, heart, and liver damage, and increase the risk of breast cancer.

It is not just the plastic in the sea which is dangerous. Chemicals, heavy metals, pesticides and fertilizers, sewage, medicines, cleaning products, oil spills and radioactive waste are all in our seas too and all of them are harmful to wildlife and to humans. It is extremely difficult to clear up oil spills and the chemicals used to disperse the oil are also harmful. Chemicals and heavy metals can stay in the food chain for a long time. Chemicals such as polychlorinated biphenyls (PCBs) have been found in the sea. Thirty thousand tonnes of PCBs are used in countries bordering the North Sea. They were banned in this country forty years ago but can pollute the land and sea for decades. These chemicals affect reproduction and resistance to diseases.

NOISE POLLUTION
Ninety percent of freight worldwide is carried by sea and combined with fishing vessels, cruise-liners, research vessels, speedboats and pleasure craft, the sea is a busy, and a noisy, place. British waters are some of the most intensive in the world with an average of 400,000 ships per year. There is also noise pollution from oil and gas exploration who use ships equipped with high-powered air guns to fire compressed air into the seabed every ten-twelve seconds. In addition, there is noise from the military testing sonar and seismic tests. The effect of all this is that the sea has become a much noisier place than it was a couple centuries ago. In fact, the University of Bristol estimates that the oceans are 36 times noisier than they were in the early 1800s. The problem is compounded because sound waves move faster and further through water than they do through air. As the sea continues to warm up these soundwaves will travel even faster. The noise can travel for more than 2000 miles (3219km). The increase in noise pollution has greatly affected many species who rely on echolocation to communicate, navigate, breed, and hunt. As Professor Carlos Duarte at King Abdullah University of Science and Technology in Saudi Arabia explains, "Marine animals can only see across tens of metres at most, and can smell across hundreds of metres, but they can hear across ocean basins." Professor Steve Simpson at Exeter University describes noise pollution as, "...acoustic fog..." which affects the marine animals ability to communicate and to hear properly. It is particularly affecting whales and more of them are being stranded each year. The Cetacean Strandings Programme began in 1913. Since then, more than 12,000 whales and dolphins have been washed up along the British coastline. It is thought that some of these are caused by military sonar. Loud noises affect the hairs in the marine animals' ears, making them disorientated and unable to communicate properly. It also affects fish, who hear at low frequencies, making it difficult for them to navigate and making them abandon their habitats.

OVERFISHING
The demand for food from the sea has increased in recent decades and this, combined with larger, better equipped fishing fleets, has led to an

increase in commercial fishing. This is combined with an increase in illegal fishing (28% of fishing worldwide is illegal) and for an increased demand from countries such as China for marine species, such as sharks, which they use in traditional medicine. This has resulted in a third of fish stocks being overfished and the loss of as many as 90% of large predatory fish in the sea. It has also led to a huge decline in seabirds which have declined by 66% in the last 60 years.

Ten percent of all the fish caught is dead by-catch which is thrown back into the sea. Shrimp fishing is the worst offender, accounting for two-thirds of the world's discarded fish. In some cases, twenty tons of shrimp are discarded for every ton of shrimp caught. If this continues by 2048 the sea could be virtually empty of seafood.

CLIMATE CHANGE
Since the start of the Industrial Revolution the oceans have absorbed 50% of our carbon emissions and have stored up to 90% of the heat trapped on earth from greenhouse gas emissions. This has had a catastrophic effect on the oceans. It has meant that the sea is warmer. The temperature in the North Sea has risen by one degree in a quarter of a century. This might not seem like much, but it has a huge impact on the species living in the sea. Species are moving 32 miles (50km) northwards every year. Warmer seas are causing the polar ice caps to melt. In the Arctic about 14,000 tonnes of freshwater are emptying into the sea every second. This affects the salt levels in the sea which in turn affects the ocean currents which we rely on for our weather and through the rain they produce, for our freshwater. If the planet continues to warm up the circulation of the currents could stop, making the sea stagnate. This is already happening in the Gulf of Thailand. In the North Atlantic major currents are already slowing down dramatically, which has a devastating effect on habitats and our weather. Already the United Kingdom and other countries around the world are seeing an increased frequency in extreme weather conditions.

The absorption of the carbon and the heat from the greenhouse gasses has also turned the seawater more acidic. The oceans are acidifying faster than they have done in the last 300 million years. By the end of this century, scientists believe that the seawater could be 150% more acidic.

This means a loss of calcium carbonate which is crucial for the formation of reefs and corals and is also what shellfish, bivalves, crustaceans, and other sea animals with shells need in order to create and strengthen their shells. These animals are an important part of the food chain so if there are less of them, it leads to a decrease in other marine species.

The position regarding the effect of climate change on phytoplankton growth is unclear. A Canadian study which stated that phytoplankton had decreased by 40% since the 1950s and were decreasing by 1% per year has been strongly debated by other scientists. In 2015 two of the same scientists in that study found that phytoplankton was increasing in-shore but declining in open sea due to an increase in temperatures and nutrients. Dr Katherina Petrou at the University of Technology in Sydney asserts that there are regional variations. However, as temperatures and nutrients continue to increase, she believes there will be a decrease in phytoplankton. American scientists state that phytoplankton has decreased in the North Atlantic by 10% since the Industrial Revolution. They state that the Atlantic Meridional Overturning Circulation (A.M.O.C), which mixes layers of deep ocean, allowing an exchange of nutrients, has weakened, probably due to climate change. Combined with more ice in the sea, leading to a less salty ocean, this means that nutrients are not getting to the surface of the sea which is detrimental to the phytoplankton because it eats the nutrients.

Other scientists believe that phytoplankton is increasing. Scientists at Stanford University have discovered that phytoplankton in the Arctic has increased by 57% in twenty years. This might appear to be a good thing, but the scientists found that an influx of nutrients in the Arctic Sea was causing a, 'thickening algae soup' which could have an impact on other species. Other species of algae, such as cyanobacteria, prefer warmer temperatures and more carbon in the sea and so they might produce more harmful toxic blooms which will increase the toxins in the sea and affect ecosystems. It might also allow for changes to the bacteria found in the oceans and might allow some human pathogens to flourish. There is also concern that higher temperatures might affect the time at which phytoplankton bloom. Presently, there is a delay between when phytoplankton blooms and when zooplankton blooms. The zooplankton eats the phytoplankton and, in turn, is consumed by many marine species. Since 2010 phytoplankton blooms fifty days earlier in the Spring. As temperatures increase, the amount of zooplankton will increase, reducing the blooms between phyto- and zooplanktons and affecting the food chain which so many species are dependent on.

Lennart Bach at the University of Tasmania has studied one of the most prevalent types of phytoplankton, diatoms, which are responsible for 50% of organic matter in the sea and 20% of the air that we breathe. He thought that they would benefit from a more acidic sea but accepted there were many variations. However, a study by Dr Petrou thought that diatoms would be affected by increased acidification of the seas because it would affect its ability to make the silica required to make its cell walls. This is serious because the cell walls are glass-like, dense structures which means that the diatoms sink more quickly to the seabed, compared to other phytoplankton, taking carbon with them, where it is stored for millennia. A more acidic sea reduces silification by between 35-80%, meaning less diatoms will make it to the seabed and less carbon will be sequestered by the sea.

Professor Sergei Petrovskii at the University of Leicester, believes that if the temperature continues to rise by 6% by 2100, as some scientists have predicted, phytoplankton could stop photosynthesising and thus stop producing oxygen. This is exceedingly bad news because not only do marine inhabitants rely on it as a food source but humans, and indeed all life on earth, relies on it for most of their oxygen. This would lead to mass extinctions including, possibly, our own.

However, scientists at the University of California are much more optimistic. They said that other studies are based on earth system models which are based on laboratory studies of phytoplankton. They dispute the accuracy of this method. Dr Adam Martiny explains that scientists traditionally measure the amount of chlorophyll in the water to determine the amount of plankton. However, the problem with this is that there is considerably less chlorophyll at hotter lower- latitude areas nearer to the Equator, as opposed to cooler regions. Therefore, he argues that laboratory tests give false results. He and his fellow researchers analysed samples from 10,000 different locations to create a global synthesis of the plankton which grows in warmer regions and found that 80-90% of the plankton in these regions are picophytoplankton, which are very small plankton. They built global maps and compared the quantity of biomass along the gradient of temperature and then used a machine learning analysis to determine the difference between now and 2100 and concluded that there would be an increase of 10-20% plankton biomass. Adam Martiny explains that "...when plankton die- especially these small species- they sit around for a while longer, and maybe with higher temperatures other phytoplankton can easily degrade them and recycle the nutrients back to build new biomass."

Whilst there is a big difference in scientists' findings regarding phytoplankton, I think it is something which we should be concerned about.

HUNTING
Although commercial whaling is banned by many countries Norway, Japan and Iceland continue to commercially hunt whales. Japan alone has killed about 400 whales in the last couple of years. It is believed that the mass slaughter of whales during the 19th and 20th century might have accelerated the effects of climate change because whales sequester a lot of carbon.

There has also been an increased demand for shark and ray meat leading to more than half the species becoming critically endangered in the last 40 years.

No-where else on earth can the effect of human activity on nature and its possible consequences be seen more than in the oceans. In summary, if we continue burning fossil fuels, dumping plastics and other pollution into our seas, and destroying valuable habitats and species, then the future does not look good, not only for the inhabitants in and around our seas but for humanity itself.

What is being done to help the oceans.

The World Wildlife Fund describe the oceans as a," planetary superpower." It can sequester carbon, provide energy and food, regulates our weather systems, and provides employment for millions of people. Doing everything possible to save our seas seems a no-brainer to me.

POLLUTION

Regarding plastic pollution globally we must reduce our dependency on plastic. Many countries have banned disposable plastics and plastic bags. In the United Kingdom there is a charge on plastic bags, which has cut their use in supermarkets by 95%, although there is concern that the government have recently given shops the option not to impose this charge and have also suspended legislation to ban plastic straws. Although the government is proposing banning single-use plastic plates and cutlery, there is no indication of when this might happen. However, they have introduced a plastic packaging tax to try and encourage greater use of recycled plastic. Hopefully this will make a difference.

Another thing which might make a difference is the introduction of carrier bags made of biodolomer, with the dominant ingredient being calcium carbonate which, amongst other things, can be found in eggshells and snail shells. Prior to this, compostable bags were made from corn starch or sugarcane and only composed at high temperatures. The new bag breaks down in ambient temperatures in less than a year and can biodegrade in a day in a commercial composter. It is also much stronger than other carrier bags meaning it can be used for two years, and it is cheap to make. The company who makes it hopes to eventually use the same material to make cups, bottles, pint glasses, and cutlery. As the material is suitable for 80% of all plastic uses it gives me a lot of hope.

Plastic producers must take greater responsibility. There should be international agreements about the handling, collection, and reuse of plastics. There are calls for governments to place high taxes on polluting plastics, and make it much cheaper and easier to recycle plastics. In the United Kingdom Surfers Against Sewage have teamed up with the British Plastic Association Federation to improve how British factories make and recycle plastic.

A new recycling plant is being set up on Teesside and hopes to start operating this year. It is hoped that it will recycle 80,000 tonnes of previously unrecyclable waste every year. New technology means that the plastics will be able to be broken down into their component parts and then re-made into plastic. This can be done numerous times. The company behind it, Mura Technology, claim that the process saves 1.5 tonnes of carbon dioxide per tonne of plastic compared to if the plastic had been incinerated instead of recycled. This could be a gamechanger for dealing with our plastic waste, and potentially the global disposal of plastic.

Major retailers should also play their part. There are many companies selling non-plastic biodegradable alternatives. However, presently the biodegradable options only take up a small section of the supermarket, with the non-biodegradable alternatives being heavily promoted. Therefore, supermarkets should do more to promote biodegradable products.

There are two billion people living in countries without proper waste disposal systems. International aid should be used to provide waste management and recycling infrastructure in these countries, according to Nina Jenson of the REV ocean project. The Circulate Initiative is going to improve waste management and recycling in five Asian countries. It is believed that this will reduce plastic waste entering the oceans by 45%. The World Economic Forum is helping organisations who have ideas for reducing plastic to scale up their business. One business they help is an African company called Madiba and Nature, which recycles plastic waste across Cameroon and uses plastic bottles to make boats. Oceanium is also funded by the Forum. They are a bio-packaging company, using seaweed to make packaging.

Eighty per cent of plastics come into the sea from the land. There needs to be better water and sewage systems on land, as well as governments being tougher on companies who pollute. The Clean Up the World project engages 40 million people in 120 countries worldwide to manually clean beaches. There are various schemes in the United Kingdom.

Some people are looking at ways of degrading plastics more quickly and safely, and ways of removing existing plastic and microplastics from the sea and coastlines. As part of a school project a Canadian teenager called Daniel Burd isolated microbial strains of two types of aerobic bacteria and used it to degrade polythene. He discovered that at a temperature of 37°C, 43% of the plastic degraded in six weeks. It was an important discovery with wide ranging implications. However, he is not the only person to have discovered that bacteria can be used degrade plastic. Professor John McGeehan at the University of Portsmouth has created an enzyme made from PETase and MHETase which is produced from bacteria which feeds on PET plastic, which is the plastic found in plastic bottles. The PETase attacks the surface of the plastic, and MHETase breaks it down further. The enzyme also breaks down PEF and polyethene furoate, two other common types of plastic, in a matter of days. Professor McGeehan and his team are trying to develop other enzymes which break down other plastics.

The Ocean Clean-Up project is an ambitious not-for-profit organisation which has developed a technique for removing rubbish from the ocean and hopes to remove 90% of the rubbish from the seas within the next twenty years. The executive, Boyan Slat, is backed by the Dutch government. Mr Slat has designed a passive collection system which is like a giant snake made up of different sections. It floats in a U-shape. Underneath the 'snake' is a 10 foot (3m) 'skirt'. The device is designed to 'shepherd' the rubbish. The rubbish is then removed from the sea and made into expensive sunglasses which are then sold to fund the project. The project has its critics who are doubtful about the amount of rubbish it will actually remove, because a lot of rubbish sinks to the bottom of the ocean, and some people worry about whether it will be a danger to wildlife, which is disputed by the organisation. Every six weeks a ship collects the rubbish and brings it back to land, so Professor Richard Lampitt of the National Oceanography Centre is concerned about the carbon cost of creating sixty of these devices and bringing the rubbish back to land. The Ocean Clean Up project say they offset their carbon emissions. It is not clear how much rubbish they have been able to remove so far.

In Plymouth they are trialling a sea bin. It is a floating bin which collects 90,000 plastic bags, 35,700 cups, and 16,500 plastic bottles a year. It is hoped that this idea can be eventually rolled out around the United Kingdom.

These schemes are great but the real problem with plastic waste are microplastics. Fortunately, scientists across the world have come up with solutions of dealing with it. Fionn Ferreira has created ferrofluid, a mix of oil and powdered rust, which acts like a magnet and removes 88% of microplastics from water samples. He hopes to incorporate his findings into existing filtration systems used by water companies. Currently, the filtration systems do not filter microplastics. He is hoping it can also be incorporated in marine engines. Dr Juan Jose Alava, who is an expert on marine ecotoxicology, is looking at how sea cucumbers and microbial comms, which are bacteria which can break down polymers in plastic, can be used to break down microplastics. In the U.S.A. Marc Ward, from the Blue Wave Initiative, uses a statically charged screen to remove microplastics from the sand on the beach. Meanwhile in Finland, the V.V.T. Technology Research Centre has created nets made of nanocellulose mesh, which can bind with microplastic as small as 0.1mm in diameter. These are all very new innovations, and it is to be hoped that they can be done on a large scale over the next few years. If so, the microplastic problem might hopefully be dealt with.

Many people are looking at alternatives to plastic which are fully biodegradable and non-toxic. Last year, at Osaka University in Japan, researchers made low cost, biodegradable plastic, made from cellulose and starch. It is hoped that it can be mass produced. James Longcroft is a Scottish inventor who has created a plastic bottle which disintegrates within three weeks if it enters salty water. He is looking for funding so he can mass produce the bottles.

There are many projects tackling the problem but what is clear to me is that governments and businesses should be funding these projects so that they can be adopted on a much larger scale, which would hopefully make a big difference.

In terms of waste from the fishing industry polluting the seas, there has been another potential breakthrough. Several companies are developing ropeless fishing gear, triggered by a timer which triggers the trap to raise to the surface. It is thought that the technology is going to be expensive, but it is surely worth doing because it is thought that it will reduce entanglement by as much as 90%.

NOISE POLLUTION
Marine traffic fell during the covid- pandemic allowing whales, in particular, to communicate properly. However, it is expected to increase again as the pandemic slowly loses its grip on the world. Currently, 85% of noise from ships comes from the propellers. To reduce the noise, the design of ships needs to be improved. Some ships are already being

designed with propellers that reduce sound pressure. Reducing the speed of vessels also helps. Just by cutting the speed from 15.6 knots to 13 knots, reduced the noise pollution in the Mediterranean by 50%. The charity I.F.A.W. suggests that ships which take measures to reduce noise pollution should receive a reduction on port fees. Re-routing shipping lanes so that they avoid sensitive ecosystems has been introduced in Canada and needs to be adopted globally.

In addition, Total, Shell, and ExxonMobil have come together to design a quieter way of carrying out seismic surveys. In 2023 they are introducing a system of using continuous lower-amplitude vibrations instead of compressed air. Not only will this be quicker, but it is also much quieter.

Meanwhile, acoustic bubble curtains are used to reduce the sound of pile driving for offshore wind farms. This method reduces acoustic energy by 95%.

OVERFISHING
The U.N. Development Programme believes that there must be a reduction in fishing by 10-15% to allow fish populations to recover. Although this unfortunately looks unlikely at the present time, things are being done to tackle overfishing. Despite covering nearly three quarters of the world's surface, only 5% of the oceans are currently protected. There is a campaign by 102 environmental charities to protect 30% of the oceans by 2030. At a recent One Planet Summit a coalition of countries, include the United Kingdom, have pledged to protect 30% of the seas. However, there is some scepticism because a similar pledge was made in 2010 and yet very little was actually done.

The idea behind the campaign is to create marine conservation zones across the world. Some damaging activities are then meant to be banned from these zones. Currently the United Kingdom, has 76 marine conservation zones, covering 25% of its waters. It is hoped that eventually there will be about 350 around the British coastline. Unfortunately, this only covers in-shore waters, within 12 miles of the coast. Whilst trawling has been banned in a handful of these zones, allowing the seabed and fish populations to recover; there are still complaints from organisations such as Greenpeace, that trawling and dredging is still taking place, sometimes illegally, in more than 97% of marine conservation zones. Greenpeace have recently sunk large boulders in the sea near Dogger Bank (an important breeding ground for fish and sand eels) and off the Brighton coast to try and prevent this. However, in some areas, these zones are making a difference. Unique and fragile reefs in Lyme Bay, off the coast of Dorset, were being severely damaged by scallop dredging. However, within three years of the zone being created, the reefs were recovering, helping various other species. Joan Edwards from Living Seas project at the Wildlife Trusts was amazed because she had it expected it to take at least two decades to recover. Within five years of Britain's first marine conservation zone been created, at Lundy Island off the coast of North Devon, there was already an increase in the number and size of lobsters as well as a noticeable increase in seabirds and seals.

Recently one of the largest marine protection zones in the world was created around the coast of the most remote archipelago in the world- Tristan da Cunha. These islands are 1750 miles (2815km) from South Africa and 2088 miles (3360km) from South America. It takes longer to sail there from Cape Town than it took Apollo 11 to reach the moon. Seven thousand kilometres square of the island's waters (an area three times the size of the United Kingdom) have joined the United Kingdom's 'Blue Belt'. This makes it the largest marine protection zone in the Atlantic and the fourth largest in the world. This is good news for the tens of millions of seabirds, seals, and cetaceans that live in the area.

However, organisations such as the World Wildlife Fund are critical of marine protection zones describing them as, "...empty words of a page..." unless they are properly monitored. The Wildlife Trust are monitoring the marine conservation zones as part of their Living Seas projects, but there needs to much more investment in monitoring these zones. The Fisheries Act also needs to be extended so that it gives protection to all the marine conservation zones and not just four of them, as it does at the present time.

There are several things which the fishing industry and consumers can do, and in some cases are doing, to prevent overfishing. There are national and local laws banning fishing in some areas and limiting when and what they can fish and the type of equipment which can be used. Fishermen are being encouraged not to fish for endangered species. The World Wildlife

Fund is working with seafood companies to encourage sustainable and environmentally friendly fishing. Catch shares give a specific portion of the year's overall catch to individual or groups of fishermen. This limits the number of fish caught, allowing fish populations chance to recover, but also gives fishermen more security.

Conservation organisations argue that trawling should be banned because it kills things indiscriminately and damages marine ecosystems. They argue that the duration of the fishing season should be shortened to allow fish more time to reproduce.

There also needs to be more done to stop illegal fishing, but that is easier said than done. However, an organisation called Sea Shepherd is arresting illegal fishermen off the coast of Gabon.

By-catch is decreasing in recent years as there is a more selective use of fishing gear and better practices by fishermen.

CLIMATE CHANGE
Only 8% of the United Kingdom's seagrass meadows remain. It has been damaged by fertilizer run-offs and has been deliberately removed by fishermen and hotel owners.

Seagrass is a bit of a misnomer because it is not a grass and is actually related to orchids and lilies. However, it has long leaves which make it look grass-like. There are 72 species of seagrass in the United Kingdom. They like shallow, brackish water where the roots (rhizomes) burrow into the sand in order to obtain nutrients and keep them securely in place. Seagrass can clone itself, allowing the rhizomes to spread across the ocean floor creating 'meadows.'

Seagrass has been described as the 'lungs of the sea' because they release oxygen and store carbon dioxide. One square metre of seagrass releases ten litres of oxygen per day. A hectare of seagrass stores 63 stone(400kg) of carbon per year. Seagrass stores 10% of the carbon in the sea. This might not sound like much, but currently seagrass only makes up 0.2% of the seafloor. Even better when the plant dies it sinks to the bottom of the ocean, burying the carbon in the sediments of the ocean bed.

Seagrass is incredibly resilient. It can survive changes in salinity, extreme temperature fluctuations, strong currents, and extreme weather.

Understandably scientists think it might play a big part in tackling climate change. It also helps the marine ecosystem in other ways. It provides nurseries for fish, with fifty species of fish relying on it. It also acts as a water filter, protects coastlines from extreme weather by reducing wave energy, and stabilises sediments through its extensive root system, thereby reducing erosion. It is even thought it might play a role in alleviating the acidification of the oceans by providing vital buffers. Dr Koweek of the Carnegie Institution for Science explains, "We are beginning to understand that some marine organisms, such as mussels, are actually able to shift the time of day which they do most of their calcification. If other organisms do the same, then even brief windows of significant ocean acidification buffering by seagrass meadows may bring substantial benefits to the organisms that live in them."

Project Seagrass aims to restore seagrass meadows around the British coastline and is currently working off the Welsh and Essex coastlines, as well as East Anglia, the Solent, and the Isle of Arran. Other projects are doing similar things. The good thing about seagrass is that it is fast growing, so if the meadows are protected hopefully there will be a big difference in a short space of time.

Seagrass is not the only aquatic plant which might be crucial in the fight against climate change. Seaweed also plays a multitude of beneficial roles. Like seagrass, it sequesters carbon dioxide. Kelp absorbs five times more carbon dioxide than land-based plants. The World Bank predicts that 500 million tons of seaweed could absorb 135 million tons of carbon, 10 million tons of nitrogen, and 15 million tons of phosphorus by 2050. That is a third of the nitrogen and phosphorus already in the oceans. By reducing the carbon dioxide in the sea, it helps reduce the acidity of the sea. Also like seagrass, seaweed protects the seafloor, improves water quality, and is an important habitat for many sea species, especially shellfish. Recently a joint study by Tel Aviv University and University of California suggests that if seaweed farms were placed in estuaries it would significantly reduce nitrogen in marine and estuarine ecosystems.

Off the Scarborough coast a firm called Seagrown is producing seaweed to be used in biodegradable plastics, cosmetics, textiles and biochemicals, as well as for consumption. Aquaculture is relatively new, but it is something that many scientists are saying should be developed further because it would be beneficial for the environment as well as being profitable. A kilometre square frame can be sunk 25 metres under the sea. It would not cause a risk to wildlife. Seaweed grows thirty times faster than land-based plants and could be cut on a ninety-day rotation. Currently, the European Union is being urged to establish a legal framework for granting licences to new and established seaweed farms.

Whales are also been seen as a key part of improving the marine ecosystem and combating climate change. They feed at depth but come up to the surface to breathe, circulating nutrients, and then when they poo they release nutrients, which phytoplankton then feed on. An International Monetary Fund report calculates that if the number of whales increased to their pre-whaling levels of about four to five million, the amount of phytoplankton in the sea would be increased. As phytoplankton absorbs more carbon than the Amazon rainforest four times over, as well as provides a large amount of the air that we breathe, this would greatly benefit the marine ecosystem and humans. The International Monetary Fund believes that even an increase of phytoplankton by 1% would capture the same amount of carbon as two billion mature trees. It is not just the phytoplankton that can store the carbon. One whale can sequester as much carbon as 30,000 trees every year. Whales continue to be useful even when they die because they take a lot of carbon to the seabed and provide mini ecosystems for other sea life. The International Monetary Fund's report stated that, "... each great whale sequesters 33 tons of carbon dioxide on average, taking that carbon out of the atmosphere for centuries." The same report claimed that if populations of baleen whales were allowed to recover, they would store the same amount of carbon equivalent to 110,000 hectares of forest.

The sea absorbs about a third of the carbon dioxide in the atmosphere, and contains more carbon than the soils, animals, plants, and atmosphere combined. Every cubic metre of the ocean contains four ounces (120g) of negatively charged bicarbonate ions, which balance with positive ions such as calcium and magnesium. Scientists are looking at artificially increasing the amount of bicarbonates in the sea in the hope that it will reduce the acidity of the oceans. By absorbing carbon, the pH levels in the oceans have been slightly decreased and this has led to acidification of the seas. If carbon emissions ended today the oceans would eventually absorb all the carbon dioxide, but the acidification would be neutralised by the dissolution of carbonate sediments. However, this would take thousands of years. Scientists believe they could greatly speed up this process. They suggest increasing the alkalinity of the sea by adding powdered rocks and minerals. This would increase the amount of calcium, magnesium, and sodium in the sea, which in turn would lead to an uptake in carbon dioxide forming bicarbonate ions, thereby sequestering more carbon. Global carbon cycle modelling believes that if the alkalinity of the sea were artificially increased the oceans could store forty billion metric tons of carbon dioxide per year with minimum environmental impact. So far though just small-scale experiments have taken place because scientists are still researching how an increase in alkalinity would affect marine ecosystems.

Restoring shingle beaches also helps fight climate change because they dissipate wave energy by absorbing 90% of the force they are exposed to, particularly useful as extreme weather conditions and storms become more frequent. This protects coastlines and reduces erosion. The best way to protect shingles beaches, according to the charity Buglife, is to leave them alone. A minimum amount needs to be done, such as preventing shrub encroachment and reducing access by vehicles and walkers, as well as ensuring unimpeded tidal patterns by removing hard defence structures which conservationists now think do more harm than good.

Sand dunes also prevent and reduce coastal flooding. The National Trust, Plantlife, and Natural England are restoring sand dunes by removing scrub, improving plant diversity.

Saltmarshes also protect coastlines by reducing erosion and storm surges and increasing the height of the coast by trapping silt. They also sequester carbon, filter water, and provide important habitats for birds, plants, fish, and invertebrates. However, globally half of saltmarshes have become degraded, mainly because of drainage, overdevelopment, pollution, and erosion. Currently in the United Kingdom saltmarshes are also being restored. The Wallasea Island project in Essex is the largest saltmarsh restoration project in Europe. It is using soil from the Crossrail project to create 1500 acres of intertidal habitats. As it is a site of national and international importance this is going to hugely benefit nature and will sequester more carbon.

A British company has recently made the world's first artificial 'digital' coral reef, 394 ft (120m) of cojoined arches. By using electrolysis, limestone forms on the structure, encouraging a coral reef to grow in a fraction of the time it would normally take to occur naturally. These structures can be placed around coastlines, helping to reduce coastal erosion. Currently there is one off the coast of Mexico which has already made a big difference to erosion in the area. As well as reducing wave energy and erosion, the new reefs will provide important habitats. It is hoped that the technology can be tweaked so it can also be used in colder seas, such as around the United Kingdom, and used to encourage beds of molluscs to form reefs of the eroding coastlines of places such as East Anglia.

HUNTING

Scientists hope that the days of commercial whale hunting are numbered. Only three countries currently hunt whales- Norway, Iceland, and Japan- and Iceland have not hunted any whales since 2019. Pressure is being put on these countries to stop whaling by trying to make it a condition of trade deals and trying to persuade people to stop eating whale meat. However, it does not look as if Norway or Japan will stop hunting in the near future. There has been a bit more success with a campaign to stop sharks and rays being hunted. One hundred and two countries have strengthened their laws regarding trading of shark and ray meat. It has been banned by the United States and other countries, although unfortunately not by some of the worst culprits, including China and Japan.

NINE STEPS TO SAVE OUR SEAS

Professor Douglas McCauley of the University of California, Santa Barbara, proposes a plan to save our seas:

1. **Freeze the warming**- There needs to be ambitious international agreement about how to achieve carbon neutrality. This is going to require big changes if it is to be achieved and can only be done so with international agreement and co-operation.

2. **Walk the talk**- Talking about carbon neutrality is not enough, internationally governments must invest in low- carbon energy as well as improving marine energy infrastructure, such as building low emission ships.
3. **Blue Revolution-** Aquaculture has increased by 1000% which is great news, but steps must be taken to ensure it is done in the correct places with the right species.
4. **30x30-** Select the best 30% of the ocean to protect, which is going to lead to the greatest gain. New technology is urgently needed to monitor and protect these 'living assets.'
5. **The other 70%-** There needs to be better management of the fishing industry and ocean mining should be banned.
6. **U.N. Treaty-** a proposed U.N. Treaty to protect and manage the seas around Antarctica and for high seas biodiversity would help.
7. **End Plastic Pollution-** Ban single use plastic, find alternatives, improve waste management and recycling facilities.
8. **Land-** Save forests, such as the Amazon; prevent fertilizer run-offs and reform farming.
9. **Wired ocean-** More data is required. Better technology is required to detect illegal fishing and protect endangered species.

What you can do to save our seas

- Think about your plastic use and change to non-plastic, biodegradable alternatives where possible. There are many eco-friendly products and companies. It might cost a fraction more, but the environmental gain is huge.
- Lobby your local M.P. and councillors for better recycling facilities.
- Lobby the government and advise it that it must provide better monitoring for marine conservation zones and must honour the pledges it makes in international agreements. Also remind it that the environment should always be a big deciding factor in any proposed infrastructure projects.

The Wonders of the Wild Places

- Clean up your local beach. If you look on the internet or social media, you will find a local group or charity in your area.
- Be careful about the chemicals you use at home. There are many environmentally friendly products available.
- Join in the 30x 30 campaign – use the hashtag #Love30x30 on social media and spread the word.
- Adopt a whale- see the W.W.F. website, www.worldwildlife.org for further information.
- Join a wildlife organisation such as the R.S.P.B. or Wildlife Trusts. If you cannot afford to join them why not volunteer for them? All the details are at the end of this book.
- If you have a boat, think about ways of making it more environmentally friendly, such as reducing your speed.
- Have a look at the merchandise sold by the Ocean Clean-Up project at www.theoceancleanup.com and help to remove plastic from our seas.
- Switch to green energy. Firms such as Octopus get all their energy from green sources.
- Buy locally where possible to reduce the amount of freight carried by sea.
- Choose fish which has been sustainably caught.
- Take part in the Dynamic Dunescapes project. See www.dynamicdunescapes.co.uk for more information.

Urban Jungle

About 55 million people in the United Kingdom live in urban areas, as opposed to about 10 million living in rural areas. You might think that there is not a lot of urban wildlife, but you would be wrong. This chapter is divided into some of the different urban habitats.

Gardens

The Ancient Egyptians are the first people known to have had gardens enclosed by walls with trees in a row. They associated different plants and trees with different Gods so grew a variety including poppies, roses, irises, daisies, and date and pomegranate trees. Ancient Greeks were the first to plant things in containers, and to plant orchards.

Unsurprisingly, it was the Romans who introduced the concept of gardens to Britain in the first century A.D. They had kitchen gardens next to their villas and had formal gardens where they had symmetrical box hedges and gravelled paths. There were niches in the hedges where they could place statutes or urns. They also introduced many plants and vegetables to Britain including roses, irises, leeks, plums, and turnips.

Gardens were developed further during the Middle Ages when monasteries had gardens where they grew herbs for food and medicine, as well as having dovecots and fishponds. In Tudor times knot gardens were popular. This was influenced by Italian gardens and contained an intricate pattern of box and lavender hedges, viewed from a raised mound. For the first time since Roman times statutes and sundials appeared in gardens. During this period many plants including tulips, marigolds, sunflowers, horse chestnut trees, and potatoes were introduced to Britain from the Continent. The Stuarts took their inspiration from the French and wealthy Stuarts had broad avenues which swept away from their house. After the Glorious Revolution, William and Mary ruled England, and the Dutch style of garden became popular. This incorporated water features and fountains as well as topiary, bulbs, and trees in containers.

During the Georgian period people were influenced by what they saw on their Grand Tour around Europe. It was during this period that Britain developed its own style of garden through the influence of landscape designers. The most famous one was 'Capability' Lancelot Brown who was born in Northumberland in 1716 and later became head gardener at Stowe in Buckinghamshire where his employer, Lord Cobham, let him take commissions from his aristocratic friends. Before long anyone rich enough to afford a stately home wanted Brown to design their gardens and parkland, and by the 1760s he was earning £6000 per year (the equivalent of £800,000 per year today.) Even the King, George III, wanted him and he became the King's master gardener at Hampton Court in 1764. He got his nickname 'Capability' because he told his clients that their property had, ...capability for improvement." Not everyone liked his designs. The poet and satirist, Richard Owen Cambridge, said he hoped to die before Capability Brown so that," ...he could see Heaven before it was improved," by Brown.

When 'Capability' Lancelot Brown died another landscape designer, Humphrey Repton, capitalised on it and sent out circulars to the upper classes advertising himself as a landscape designer, despite at the time not having a lot of experience. He became an overnight success because of his Red Books which had explanatory texts and before and after pictures, allowing the client to see how he could transform their land. Most of these designs were based on other people, mainly Brown's, designs. Cannily, Repton charged clients to use his Red Books but did not actually do the work himself.

Due to exploration and a fascination in exotic plants, gardens changed a lot in Victorian times. Wealthy people planted arboretums and rockeries to show off their rare and exotic plants and trees. There was also a fascination in the Far East, and Japanese and Chinese gardens became popular. Then the Arts and Crafts movement transformed gardens into places where flowers were beautiful and practical and there were lots of unexpected views. A big influence on this movement was Gertrude Jekyll who created 400 gardens in Britain, Europe, and America in the later 19th, early 20th century. She spent most of her life at Munstead Wood, Godalming in Surrey where she propagated new plants, becoming so popular that by the 1880s horticulturists were all wanting to work with her. Incidentally, her brother was good friends with Robert Louis Stevenson, and it is thought that he used his name when naming his character Dr Jekyll in *Jekyll and Hyde*.

There are an estimated 24 million gardens in Great Britain. It is in our gardens where many of us engage, probably daily, with nature. According to BBC's Springwatch, 75% of us regularly feed the birds in our gardens.

One hundred and thirty-three species of birds have been spotted feeding in people's gardens. That represents more than half of the breeding birds in the United Kingdom. These species range from common birds such as robins, blackbirds, and tits, to declining birds such as sparrows, to more exotic ones, such as parakeets.

If you have seen a blackbird it might seem obvious why it got its name, but actually it was originally called an ouzel and only became a blackbird in the mid-14th century. The word bird originates from the Old English word *brid* which meant breed or brood and referred to chicks and fledglings. By the 14th century large birds were known as fowls and smaller birds were known as birds. Hence blackbirds acquired their name because they were small black birds, as opposed to the much bigger black birds, such as rooks and ravens. This distinction lasted until at least the 18th century. In his *Dictionary of the English Language* Samuel Johnson defined the word bird as meaning, "A general term for the feathered kind, a fowl. In common talk, fowl is used for the larger, and bird for smaller feathered animals." Nowadays although the term fowl still exists to determine types of birds like waterfowl, in the main it tends to be used in relation to more domesticated birds, such as guineafowl and chickens.

Robins are probably one of the best known and best loved birds in Britain. We see them on our Christmas cards or watching over us as we do the garden. However, they have only been known as robins since the 17th century, prior to that they were known as redbreasts.[79] The connection with Christmas and the reason they appear on Christmas cards is tenuous. Early Victorian postmen wore red and were known as robin redbreasts.

---

79 The British Ornithological Union did not officially recognise it as being called a robin until 1971.

The Wonders of the Wild Places

As Christmas cards began in the Victorian era many people think that robins appear on Christmas cards as a link to the postmen who were delivering them. However, there is also a story about the robin fanning a fire to keep the baby Jesus warm in his crib, as well as another legend about it watching over Christ on the Cross (both are used to explain its redbreast) so there are religious connotations surrounding this little bird.

Robins acquire their redbreasts after their first moult. Each robin has a unique breast pattern. Robins do not move far from where they are born. They are very territorial little birds and of the 75% that die before their first birthday, 10% of these are killed in territorial disputes. The majority are sadly killed by cats.

Scientists have discovered that robins can see the earth's magnetic field. Katrin Stapputt from Goethe University has discovered that it is the robin's right eye which allows it to do this. It is thought that a molecule called cryptochrome, which is found in the light sensitive cells in the robin's retina, might be responsible for this ability. Scientists think that it affects the level of sensitivity of these cells. The magnetic fields appear as patterns of light and shade which is superimposed on what a bird would normally see. These patches change as the robin moves its head, providing it with a visual compass. However, experiments have shown that they only have this ability if they have clear vision in their right eye.

Birds are also able to see ultraviolet light. In his interesting book, *The Meaning of Birds*, Simon Barnes says that "...what birds can see is beyond our experience...beyond the possibility of our imagination."

Many of us have probably tittered when talking about the various species of tits. However, these little birds were once known as titmouse, from the Old English *mose*, meaning small creature. The types of tits you are likely to see in your garden are blue, great, coal and long-tailed tits. Coal tits are black and white, and small, weighing about the same as a fifty pence coin. Great tits can be differentiated from blue tits because they are bigger and have a black 'scarf' on their yellow breast. If it is a long 'scarf' it is a male great tit. Blue tits are one of the most common birds in British gardens. They are very colourful little birds. They have big broods because so many of them are killed each year, particularly by cats.

If you live in the South- East of England (although they have been spotted in Leeds and Manchester and even as far north as Aberdeen) you might see a more exotic looking bird in your garden, a ring-necked parakeet. They have increased by 95% since the mid -1990s. Many legends exist about how these brightly coloured birds, which are native to Africa and South Asia, came to the United Kingdom. One is that they were released by Jimi Hendrix, another is that they escaped from the set of the African Queen. However, researchers, have discounted these explanations, because parakeets have been in the United Kingdom for longer than sixty years, and believe that the truth is much more prosaic. There was a craze for caged birds after the Second World War and they became hugely popular pets. However, then parrot flu came along, and many people could not afford to pay the vets bills so just released the birds into the wild. Like other parrots, ring-necked parakeets are mimics, capable of copying human speech and learning and repeating words and phrases.

As well as seeing birds in your garden you will probably hear them, especially during Spring. Normally it is the male bird who sings. They sing in order to attract a mate and to announce that it is their territory. If you have been up early enough you might have heard the wonderful cacophony of the dawn chorus. It can be heard between March and July but is at its best from late April to early June because during that time migratory birds, such as chiffchaffs, return to the United Kingdom. The dawn chorus starts about an hour before dawn, and peaks about half an hour before and after it gets light. There is a strict order with robins and blackbirds starting proceedings. Birds with the largest eyes absorb more light so they tend to be the ones which sing in the morning. However artificial light can affect when birds start to sing so I sometimes hear robins and blackbirds singing at night. Sound travels twenty times faster in cold, still air, which is why the sound seems louder than at other times of the day. About twenty minutes after the robin and blackbird, birds like warblers and wrens join in. Wrens have the loudest and most complex songs, singing at 740 notes per minute although our ears cannot detect the frequencies of many of the notes. The last to join the chorus are the birds with smaller eyes, like sparrows, great and blue tits, and finches. Scientists have discovered that birdsong changes over time. Blackcaps add more notes as they get older.

The Wonders of the Wild Places

Birds use a forked organ called a syrinx to make sounds. Some birds can sing two notes at the same time. Birds breathe differently to humans, so they can make long, loud continuous sounds. Birds make calls as well as singing. A call tends to be monosyllabic and is normally a communication of a simple idea, such as warning of danger. Songs are more elaborate and complex. Singing takes a lot of energy. Female birds can tell how fit and strong a male is by his song. Not all birds sing. For example, crows and rooks call rather than sing. Birds make a variety of sounds. Great tits sing about seventy different songs to communicate different things. Each great tit has a repertoire of about eight songs although the easiest one to listen out for sounds like *teacher, teacher.*

Urban birds sound differently to country birds. Urban birds have louder songs. Rupert Marshall, at the University of Aberystwyth, discovered that urban great tits and blackbirds sing at a higher pitch than their rural cousins. Some urban birds have adapted when the sing, in order that they are not drowned out by the noise of rush hour traffic, others have adapted the frequency of their song so they can be heard over the traffic noise. This is a problem because low level frequencies are less likely to be degraded (broken up) by things like vegetation. However, these frequencies become more degraded if they are sung from less than three feet (1m) off the ground. The lockdowns, because of the covid-19 pandemic, had a positive effect on wildlife in general and the dawn chorus was noticeably louder and clearer. Traffic noise was reduced by about 90% during Spring 2020, greatly reducing noise pollution.

Scientists have discovered that young male birds learn how to sing from their dad, or if he is not around, an older male tutor. Like humans they learn by listening, imitating, and practising. They have a special part of their brain called the High Vocal Centre which helps the song to become imprinted in their brains. It takes about three months before they have mastered their song. It will not be identical to their dad's or tutor's because birds sing to identify as a particular species, but also to distinguish themselves from their male competitors. Individual birds have their individual repertoires and so there is a large variation within species. However, birds can recognise their species-specific songs even if they have not been taught it during the first three months of life. In the 1950s William Thorpe did pioneering research into how birds learn to sing. He found that chaffinches which had been raised without an older male, developed abnormal songs. However, when they were played recordings of chaffinches singing, they recognised it as being their species and learned to sing specific – species songs. This shows that birds have an innate ability to discriminate between their species – specific songs and other songs. Their heart rate increases when they hear another bird of the same species singing.

After the birds have found a mate, they will make their nests. Like their songs there is also a great variation in their nests. With some species it is created by the male, in other cases the female, and sometimes both birds build it together. Different species use different techniques and different materials. One of my favourite birds are long-tailed tits which live in extended family groups with family members helping to look after the young. Long-tailed tits might win the prize for the cosiest nest. They use moss, hair, wool, feathers, and spider's webs, which allows the nest to expand as the family gets bigger. Studies have shown that long-tailed tits will fly between 600-700 miles (966-1127 km) collecting all the materials they need for their nests.

Birds use their beaks to build the nest. Songbirds, like blackbirds and robins, make classic shaped nests. They use their beak to carefully weave the materials. The bird will turn in the inside of the nest to compact the material and make the inside of the nest smooth and cosy.

It is no wonder that some birds chose just to utilise a space or an old nest that someone else has made. Many tits and owls will just nest in a hole in a tree, and birds like starlings and sparrows nest in the eaves of houses. I have house sparrows nesting in my eaves. It can be a bit noisy and sometimes it sounds as if they are wearing clogs, but I really like knowing they are there especially as they are a species which has dramatically declined in recent decades. Sparrows and starlings will still line their nests. The house sparrows living in my house like to use my neighbour's pamprass grass to create a soft and warm lining. Nest building takes a lot of energy so birds like sparrows, which do not take a long time to make a nest, can breed more quickly.

According to Springwatch, 90% of Brits have plants which provide nectar in our gardens. One of these is lavender which was introduced to the United Kingdom by the Romans who valued the plant so much that it used to cost a month's wages to buy one. The name either comes from the Latin word *livere* meaning 'to be blueish', or the Italian word *lavande* meaning to wash. In the Middle Ages it was regarded as a wonder plant because of its medical properties, including aiding digestion, and soothing stings and bites. During World War One it was used to treat burns. Lavender has been shown to have antibacterial properties.

Another plant which pollinators like are sunflowers, which originate from the Americas, and were an emblem of Christian devotion and fidelity which is why they can often be seen in 16th century wedding portraits. When they are at the bud stage, the sunflower rotates during the day, following the movement of the sun from East to West. Then at night they turn back to the East again ready for morning. This is called heliotropism. The French word for a sunflower is *tournesol* which literally means 'turned sun', and their Latin name *helianthus* refer to this ability. By doing this the sunflower gives itself the best chance of converting as much sunlight throughout the day in order to convert it into energy. They need between six and eight hours of sunlight per day. Once the flower has bloomed they stop following the sun because they are too heavy. At this stage the sunflower permanently faces East. This allows it to be warmed up by the morning sun, which makes it stronger and bigger, but also attracts pollinators. The sunflower might look like one big flower, but if you have a close look at it you will notice it is actually made up of about two thousand smaller flowers called florets, which are laden with nectar and pollen, perfect for bees and other pollinators.

Sunflowers ability to use heliotropism allows them to grow higher than a lot of other plants, which also helps them absorb the maximum amount of sunlight. Normally they grow up to twelve feet (4m) in six months, but the world record sunflower, grown by Hans-Peter Schiffer in Germany, was a whopping 30ft 1 inch (9m) tall. It was so tall that the fire brigade where used to measure it, and it required scaffolding to keep it up.

Sunflowers have existed in the Americas for at least 5000 years and since that time the oil has been used for medicine, food, and dye. They were brought to Europe by Spanish conquistadors in the 16th century.

The Wonders of the Wild Places

Probably the most amazing ability of sunflowers is their ability to remove radiation and dangerous metals from contaminated soil. Catie Kitrinos at the University of Virginia explains that, "Sunflowers are able to take heavy metals from contaminated soil in a way that is completely natural and un-harmful to the soil and its surrounding ecosystems." Other plants, such as cactus and aloe vera, also have this ability, but sunflowers are particularly useful because of their size. After the Chernobyl disaster in 1986, which emitted fifty tons of radioactive material into the atmosphere, fields of sunflowers were grown in the area. They absorbed the radioactive isotope cesium from the contaminated soil and into their stems, flowers, and leaves. The sunflowers were then harvested and safely disposed of by a process called pyrolysis, which is when the carbon is burned off and then the remaining radioactive metals are safely stored. However, the same technique was used after the Fukushima nuclear disaster in Japan and was much less successful, only absorbing 0.2% of the radioactive material. Scientists have discovered that not all sunflowers have the ability and that it is not just a difference between different species of sunflowers, but within individuals of the same species. Currently, they do not understand why this is. Catie Kitrinos and her colleague Assistant Professor Benjamin Blackman are trying to find out and are experimenting with whether they can create a species which can reliably be used to decontaminate land.

Two flowers which are related to sunflowers, but are far less popular with gardeners, are daisies and dandelions. These are generally regarded as weeds, but there is no such thing as a weed. Just because a flower is growing somewhere which a human might think is inconvenient does not make it any less beautiful or useful. Habitats are a human concept; plants grow wherever the conditions are right. Once you banish the w-word from your vocabulary you start to see plants in a totally different light.

There are many species of daisies, and they live all over the world, except in Antarctica, because they can grow in wet and dry conditions. They get their name from the Old English *daes eag* meaning *day's eye* because they open at dawn and close at dusk. The French name for them is *marguerite* which is derived from the Greek word *margaron* which means *pearl*. This is because the delicate white petals are said to look like pearls. Incidentally, daisies are a composite flower with the yellow 'eye' being one flower and the pearl-like petals another. The herbalist Nicholas Culpepper referred to it as *wound herb* because it is said to aid healing. This plant has wide ranging medicinal uses, from slowing bleeding and aiding indigestion, to helping with arthritis and stomach ulcers, to fighting infections and helping with menstrual problems. No wonder it was Geoffrey Chaucer's favourite flower. Daisies were associated with the Virgin Mary and represent purity and innocence, which is maybe why they are strung together in chains and hung around young girls' necks.

Dandelions get their name from the French *dent de lion* meaning lion's tooth, because of the shape of their leaves. Another name for them was wet-the-bed probably because they are a diuretic. Dandelions have been used in Chinese medicine for thousands of years and are a useful plant because every part can be used for food, medicine, and dye. The leaves are very high in minerals, vitamins, and antioxidants, so eat them in a salad. Like daisies and sunflowers, the flowers open in the morning and close in the evening. Specifically, they open at 5am and close at 8pm so were known as the shepherd's clock.

The dandelion is the only flower which is said to represent the sun (the flower), moon (the head when it has gone to seed) and the stars (the seeds.) The seeds are disbursed by the wind and can travel five miles (8km.)

Dandelions are associated with healing, radiance, transformation, clarity, and strength. There are numerous superstitions and rhymes associated with them. People are encouraged to make a wish when they blow on a seed head. The number of seeds left on the head after one blow will either tell you how many children you will have or how many years you have left to live, depending on who you listen to. My sister and I used to say, "What time is it Mr Wolf?" and then blow on the seed head. The number of attempts it would take us to blow off all the seeds denoted the current time. I cannot attest its accuracy and suggest that looking at a clock is a better way of telling the time.

Next time you get frustrated by the dandelions in your lawn remember the words of one of possibly the best nature poet, John Clare, in his poem *A Rhapsody*, "...we almost think they are gold as we pass or fallen stars in a sea of green grass."

Dandelions and daisies are both good sources of nectar and pollen.

Roses are also popular with pollinators, as well as gardeners. Roses have been around for at least 35 million years and the oldest living rose, which can be found at Hildestein Cathedral, is thought to be a thousand years old. Modern cultivated roses are all related from the first hybrid rose *La France* created in 1867.

Roses have been associated with love and beauty since at least Ancient Greece. The first paintings of roses appear in Ancient Greek frescos. In Roman times newly married couples were crowned with crowns of roses.

In Greek mythology the Goddess of Love, Aphrodite, is said to have created the first rose from her tears and the blood of her lover, Adonis. The Romans copied the story, changing Aphrodite into Venus, their Goddess of Love. Venus's son was Cupid, who is still associated with love. He supposedly fired arrows into a garden. The arrow caused the roses to develop thorns, which Venus then pricked herself on, with her blood turning the roses red. Cupid is also said to have offered a rose to the God of Silence as a bribe to stop him giving details of Venus's various amorous escapades. Thereby, in Roman culture roses also became a symbol of secrecy. Roman dining rooms were covered with murals of roses to remind guests that whatever was discussed around the table must not be repeated once the guest had left the room.

In Medieval times the rose became associated with the Virgin Mary. Each petal symbolised a different joy of Mary- the annunciation, nativity, resurrection, ascension, and assumption. In Medieval paintings Mary was often depicted in a rose garden. Many churches and monasteries had rose windows, and in the 13th century rosary beads were first created to help Catholics recite their prayers.

Historians might be familiar with the turbulent period in English history when there was a power struggle between the houses of York and Lancashire in the mid to late 15th century, which was finally concluded at the battle of Bosworth in 1485 with the death of Richard III. This period is known as the War of the Roses. It has only been called that since the 19th century when the Scottish author, Sir Walter Scott, coined it in his novel *Anne of Geierstein*.

Roses have also been associated with socialism since the 19th century. Following the French Revolution socialists wanted the red flag, another symbol of socialism, to be the French national flag but Republicans decided on the tricolour instead. However, in order to pacify the Socialists, the provisional government stated that politicians could wear a red rosette. Then in 1878 Chancellor Bismarck passed anti-socialist laws in Germany and anyone seen wearing a red ribbon, the sign of socialism, was arrested and imprisoned, so socialists began wearing red roses instead. Many of they were exiled so the practice of socialists wearing a red rose as a symbol of their beliefs spread across Europe and North America, and by 1910 it was universally accepted as the symbol of socialism but has just been used by the British Labour Party as their symbol since the 1980s.

Well planted gardens can be a great place for pollinators. Bees need all the help they can get at the moment. Since 1900 the United Kingdom has lost thirteen species of bee and 35 more face extinction. It is not just a problem in this country. Across Europe, 10% of wild bees face extinction. Unfortunately, the bees currently face many foes. Climate change affects when the flowers they pollinate bloom. Invasive species such as Asian hornets and small hive beetles, as well as pests, such as the varroa destructor, which sucks the blood from the honeybee, and fungal diseases are decimating hives. Pesticides impair navigation and reproduction. Then, as if that were not enough, they must put up with pollution and habitat loss as well.

Why does it matter if we lose some of, what Geoffrey Chaucer first coined, our busy bees? We need them to pollinate plants and crops. There are other pollinators too, but every one of the 270 species of bee in the United Kingdom has an important ecological role. There are around 24 species of bumblebees, 240 solitary bee species and one species of honeybee in the United Kingdom. Three of the most common British bees are the red mason bee, which can be seen from April onwards and is an important pollinator of fruit trees; the common carder bee, which can be seen in Spring and Summer, they weave leaf litter into nests, and have long tongues which they use to remove nectar from bluebells, bugle, and primroses; hairy-footed flower bee, who gather pollen on their hairy legs.

One of the types of bee which people are probably most familiar with is the honeybee. They have existed for 150 million years. There are between 20,000-80,000 honeybees in a hive. There is an extremely strict hierarchy in the hive. There are male drones and female infertile workers who are presided over by a Queen. The worker bees only live for up to about six weeks and make, on average, a twelfth of a teaspoon of honey during their lifetime. They convert the nectar, which they have collected from plants, into honey by adding glandular secretions. Honey is an extremely useful, as well as tasty, substance which is a natural antiseptic and antioxidant. It lasts an exceptionally long time. In 2012 some 5500-year- old honey was found by archaeologists in Georgia.

The Queen honeybee only flies once during her lifetime to mate with the drones, who are stingless males, of another hive. She must not enter the other hive though or she would be killed. The drones only job is to mate with the Queen. Most of the drones will die after mating but the ones which do survive have a cruel fate. Once winter comes there is not enough food to feed everyone in the hive, so the remaining drones are evicted and starve to death.

The honeybee Queen gathers enough sperm during her maiden flight to make eggs for the rest of her life.

Until recently it was thought that the Queen controlled the other bees in the hive. Although she does emit pheromones, which act to unify the hive as well as to inform individual workers of their tasks, it is the much smaller infertile female worker bees, who make up most of the hive, who actually make all the decisions. According to the needs of the hive they decided how many fertilized (female) eggs and unfertilized(male) eggs the Queen lays. The worker bees produce wax from glands in their stomachs to produce the cells in which the Queen lays her eggs and where the larva will grow. The Queen can lay two thousand eggs per day and about a million eggs during her lifetime.

As the name suggests it is worker bee who does all the work in the hive; from creating the cells and keeping them clean, feeding the larva and the Queen, foraging for pollen and nectar, to keeping the hive clean and well ventilated, and repelling invaders. Worker larva must be fed worker jelly by the workers about 800 times a day. After about eight days they spin a cocoon and enter the pupal stage, before emerging three weeks later as worker bees. Whilst in the larva stage the temperature of the brooding chamber must be kept at a specific temperature. if it becomes too hot the worker bees collect water and deposit it around the chamber. The water evaporates, thus cooling the chamber. If it is too cold the workers cluster in order to generate body heat.

The atmosphere in the hive as the Queen gets to the end of her life is like a mixture of a soap opera and Hamlet. With honeybees it is the workers who chose their new Queen. When the Queen's egg production reduces, she emits less pheromones, or she is sick, or dies, the worker bees will put some of her eggs in a special cell, where they feed them with royal jelly, which is mixture of protein, water, and sugar, produced by the workers from a part of their throat called the hypopharynx. The royal jelly allows the larva to grow faster and ensures they will survive longer than their siblings. It used to be thought that royal jelly made the new Queen bee fertile, but it is now thought that it is what the young Princess does not eat which ensures her fertility, not what she does eat. The young Princesses are fed royal jelly exclusively so do not eat honey and pollen like the workers. Scientists believe that the honey and pollen contain naturally occurring chemicals which cause the female workers to become infertile.

If one new Queen emerges she will destroy all the other Princesses' cells before they have time to emerge. However, if more than one Queen emerges at once there is literally a fight to the death because in honeybee hives there can only ever be one Queen. If the old Queen is still alive at the time the new one emerges, she will leave the hive, with some of the workers, and create a new colony elsewhere. This is known as swarming. However, before the old Queen leaves she must be fed less by the workers in order that she is light enough to fly.

Bumblebees also live in hierarchical colonies, presided over by a Queen, although their colonies can survive without a Queen. A bumblebee Queen does not live as long as a honeybee Queen. She only lays about two hundred eggs during her lifetime. She emits a hormone which ensures the other female bees are infertile. Occasionally a worker bee can block the hormone and becomes fertile. When this happens there is a fight to the death to determine who the Queen should be. The fertilized eggs which the Queen bumblebee lays become female worker bees and the unfertilized eggs become male drones. She only lays unfertilized eggs towards the end of her life. This is the signal to the worker bees that she will have one more batch of fertilized eggs and that they should be raised as potential Queens rather than workers.

Bees can fly at 15 mph and can fly for about seven miles (11km). They beat their wings two hundred times a second. They have five eyes, three (simple) ocelli on the top of their head and two compound eyes. The eyes have different functions. Although scientists are still researching bees' ocelli, they believe that they help with flight stabilisation, navigation, and orientation. The compound eyes help the bee to see colour, detect even the tiniest movements and judge distance. Bees always fly home in a straight line which is where the phrase 'beeline' comes from.

Recent research at Simon Fraser University in Vancouver has discovered that, like pigeons, honeybees have a ferromagnetic material called magnetite inside them which helps them detect the earth's magnetic field. Professor Randolf Menzel, from the Free University of Berlin, has discovered that bees also use the sun to help them navigate. They make an internal map of the landscape, using natural landmarks, and memorise flight paths. Bees can see polarized light which enables them to 'see 'the sun even on a cloudy day. By being able to see in polarised light and make internal maps, they create an internal G.P.S. system in front of their eyes.

As well as having an internal solar compass, honeybees also have an internal clock, which helps them keep track of how far they have flown from the hive. She can work out how far the sun has moved during her journey. Once she has collected the nectar or pollen she flies back to the hive, keeping the sun at the same angle. When she returns to the hive she performs a dance, known as a waggle dance, which provides her sisters with all sorts of information including the position of the sun in relation to the flower as it is at the present time, as opposed to when she discovered it. Bees are the only animal, other than humans, who can communicate in symbolic language and about things which are not present.

This is one of the reasons why bees are considered to be highly intelligent animals. Although their brain is about a millimetre across, about the size of a poppy seed, and they only have a million neurons, as opposed to the approximately 100 billion neurons found in a human brain, bees can understand patterns and symmetry (two things associated with high intelligence) as well as perform basic maths. They can count to five and even understand the concept of zero. They can also do basic sums. Children learn to recognise symbols such as + and −. This requires use of short- and long-term memory. The short-term memory helps them to manage the numerical values and the long-term memory helps them to remember the rules. Scientists taught bees to recognise symbols. They had a Y-shaped maze which had two colours, blue which represented plus, and yellow which represented minus. If the bees flew one way it took them to the right answer, where they received a sugary treat. If they went the other way and got it wrong, they received a bitter tasting quinine solution. The answers kept being swapped around so the correct one was not always on the same side. The bees were shown the initial numbers and then had to fly to the correct answer. Incredibly they flew to the correct answer about two-thirds of the time. This required comprehension and memory, two things associated with individual intelligence.

Research at the University of Sheffield worked out that the bees could work out basic maths and puzzles by using visual clues. They taught the bees to recognise different shapes. They then taught some of them that if they touched a placard with a greater number of shapes on it, they would get a treat, and others that if they chose a placard with fewer number of shapes on it, they would get a treat. All the bees chose the right placard. They then had two placards with the same number of shapes on each, but on one placard the sides of the shape were slightly longer than on the other placard. This time there was no reward. Scientists thought that the bees would head equally to the placards but were surprised to discover that the bees who had been trained to go to the greater number of shapes flew to the placard with the shapes with the longer sides, whereas the bees who had been trained to fly to the placard with the fewer number of shapes flew to the other placard.

Similar research was carried out at the R.M.I.T. university in Melbourne, Australia. They discovered that bees could recognise that two could be expressed as two pictures of a tree, two pictures of bananas, or two hats. This indicated that they understood the concept of the number and also recognised the different symbols.

Bees possess a gene called egr, which is associated with learning in other vertebrates. This gene is activated if they are in unfamiliar surroundings. When the worker bee first leaves the hive and goes on her first few foraging trips the gene is activated in order that she can learn to navigate more quickly.

For a long time, it was thought that bees just had hive intelligence but researchers at Queen Mary University in London have shown that individually bees have the ability and the intelligence not just to learn but also to adapt and improve. They devised an experiment which encouraged bumblebees to move a ball into the centre of a platform in return for a sugary treat. They started by showing the bees how to move the ball. With some of the bees the scientists moved it with a magnet and with others they used a plastic bee on a stick, which they used to push the ball around. The bees learned to move the ball themselves. However, they then trained bees which had not seen how it was done, to move the ball. The bee pupils did not just copy what their instructors had shown them but improved on it by moving a ball closer to them, rather than further away as they had been shown. They have since taken the research even further and have taught the bees to play football.

Bees cannot only do maths they can also 'talk'. Dr Martin Bastick at Nottingham Trent University is writing the first dictionary of the bee language. Bees produce these 'words' by vibrating their wings at different speeds. Bees have four wings, two on each side which lock together when in flight. By vibrating the wings at different speeds, they produce different sounds which Dr Bastick has been able to detect by using very specialised equipment. Princesses make a quacking sound, which means they want to leave their cells. Meanwhile the Queen roams around the hive making a tooting noise which manipulates the physiology of the female worker bees, ensuring they keep the Princesses captive, because if they escaped, they would fight to the death, destroying the future of the colony. There are many other simple sounds which convey different meanings to the workers. The bees themselves cannot hear the sounds because they do not have eardrums, but they can detect the vibrations and know what different ones mean.

Bees have a fantastic sense of smell, with 170 odour receptors (although not as well as humans who have one trillion odour receptors). Research at the University of Bristol has shown that bumblebees have smelly feet. They leave smelly footprints on a flower. This enables another bumblebee to distinguish between their own scent and that of a relative or stranger, in order to determine whether someone has already visited the flower and probably taken the nectar or pollen.

A bee can only use its sting once. Its sting consists of a two darts with jagged edges. It is normally kept in a sheath. To sting someone, or more likely something, the bee thrusts out of the sheath, which has poison flowing through it, and then brings out the darts. The jagged edges make it difficult for the bee to withdraw its sting, so they normally fatally wound themselves when they use their sting. For this reason, they only use it as a last resort.

Unlike bees, wasps can, and often do, sting more than once. Only the female wasps sting. Her sting is a modified ovipositor, originally used to lay eggs but over time it has evolved as a defensive mechanism. The sting pierces the skin and contains venom producing glands, so the venom is injected into the victim. However, it does take a lot of effort to sting someone or something so wasps only do it in defence. Most wasp species do not sting, but the most common wasp species in Britain, called predictably the common wasp, does. As someone who is terrified of wasps, I am perturbed to learn that when a wasp stings, or is killed, other wasps will be attracted by a chemical in its venom and will head angrily in your direction. The best thing to do when a wasp lands on you is to keep calm, although I know from personal experience that this is easier said than done.

Wasps are related to bees and ants. There are more than 120 species in the world and there are wasps on every continent except Antarctica. They come in many different colours and vary in size. In the United Kingdom there are about nine thousand species of wasps. Like bees and ants, they go through a complete metamorphosis in their lifetimes, starting as an egg, then becoming larva, then a pupa (which cannot move or feed) and eventually emerging from its pupal skin as an adult wasp. They broadly fall into three categories- solitary, social, and parasitic. Only nine species of wasps in the United Kingdom are social wasps. The largest one is the European hornet, which is two inches (5cm) long.

The Wonders of the Wild Places

All adult solitary wasps are fertile. However social wasps are very much like bees and ants because they live in large colonies and have more than one egg-laying Queen, female workers, and male drones. At the end of the Summer the Queen(s) mate and then remain in a torpor during the Winter before emerging in the Spring. Their first job is to build a nest. They then lay eggs, who become sterile female workers, who then run the colony. In late Summer some of the eggs develop into male drones, whose only function is to mate. Also, in late Summer the Queen(s) will lay fertile female eggs, which develop into next year's Queens. The workers feed potential new Queens nectar-rich food. The wasps we see in the Summer as we are trying to have a picnic are worker wasps.

Social wasps use chemicals to identify nestmates, and to send information and warnings to other members of the colony. The *Polistes fuscatus* wasp can recognise other wasps by their individual facial patterns. This is the first time that scientists have noted an insect with this ability.

Wasps can look like bees, but most wasps have a pointed lower abdomen and narrower 'waists'. They are not hairy like bees. Wasps do pollinate, but not as much as bees. However, some plants, such as figs, rely solely on wasps to pollinate them. However, wasps do play a part in controlling pests, such as aphids. One worker wasp can collect more than a hundred aphids per day.

Wasp numbers naturally fluctuate in two, and possibly, seven-year cycles. These cycles are driven by the wasps' biology, but also by the weather. Wasps like cold Winters because the Queens remain in a torpor for longer and emerge later in the Spring when there is more food available. However, there is evidence that mild Winters are also beneficial to wasps because their colonies do not die and continue to grow. This is already a problem in Australia and New Zealand.

By far, most wasps are parasitic wasps. There are about 100,000 species of parasitic wasps, although biologists think there could be as many as two million in the world. Scientists believe that there are more different species of parasitic wasp than any other animal. Six thousand species of parasitic wasp are known to live in the United Kingdom. Almost every insect 'pest' species has at least one species of parasitic wasps who prey on it.

The story of parasitic wasps reads like a horror story. Most species lay their eggs on or inside their victim. The *Hyposoter horticola* parasitic wasp preys on the caterpillar of the Glanville fritillary butterfly. The wasps lay their eggs within the butterfly's eggs. When the caterpillar is almost fully grown, the parasitic wasp grows within them, then eat their victim from the inside before eventually bursting out of the host, who is normally still alive, before spinning a cocoon on a nearby leaf or twig and then emerging as an adult parasitic wasp. Other species of wasp wait until the caterpillar has hatched and then lay their eggs in it. Their larva feed on the caterpillar's nutrients and blood. When the larva is ready to leave its host it emits chemicals which paralyse the caterpillar. The larva then uses its saw-like teeth to eat its way out of the still-living caterpillar.

Some parasitic wasps remain inside their host and manipulate it into protecting their larvae. The ladybird parasite is an example of this. Despite emitting a poison to try and deter it, the ladybird cannot stop the parasitic wasp laying its eggs in it. When the wasp's larva have eaten its way through the ladybird's internal tissues, it bursts out of its abdomen and spins a cocoon between its legs. Thus, the ladybird effectively becomes the wasp's bodyguard. The ladybird thrashes and twitches its limbs when a predator approaches. Scientists do not know why it does it but believe that the wasp might leave venom inside the ladybird which affects how the ladybird behaves.

Some parasitic wasps prey on other wasps. Gall wasps lay their eggs in plants, which triggers abnormal growth, called galls, in which the larvae feed and grow. When they achieve adulthood the gall wasp eats its way out of the gall. However, the crypt-keeper wasp, which was only first described as recently as 2017, preys on one species of gall wasp, the *Bassettia pallida*. The crypt-keeper wasp lays its eggs in oak galls. Its larva then attacks the *Bassettia pallida's* larva. When the infected gall wasp's larva reaches the stage when they are ready to eat their way out of the gall they only chew a small hole and then stop, remaining in the hole, like a cork in a bottle. Scientists do not know yet how the crypt-keeper wasp is able to control the *Bassettia pallida* larva so that it stops at such a precise moment. However, this stops anything else harming the crypt-keeper's larva, so they safely develop. When they are ready to leave the gall, they eat their way through the head of the gall wasp who is blocking the entrance.

The Wonders of the Wild Places

Parasitoids are tiny wasps which lay their eggs on caterpillars. They emit a chemical which suppresses the caterpillar's immune system, but this chemical also changes the chemical composition of the caterpillar's saliva. When the infected caterpillar eats a plant, the plant releases molecules, which cause a distinctive aroma which acts as 'wasp-nip' to hyperparasitoid wasps, who lay their eggs on the parasitoid's offspring.

In comparison ants live quite humdrum lives, but that does not mean that they do not have any superpowers. For example, an ant can carry objects fifty times heavier than their own body weight in their jaws. Some of these powers are very extreme. The yellow goo ant in South-East Asia self-explodes if an intruder tries to enter its colony, covering the would-be predator in poisonous liquid.

There is believed to be the same amount of ants on earth, in terms of biomass, as there are people (25%) and there is thought to be 1.5 million ants for every person on earth. Over twelve thousand species of ants are known to exist. Like wasps, they live on every Continent except Antarctica. Most live in the Tropics. An acre of Amazonian rainforest is believed to contain about 3.5 million ants.

Like wasps and bees, ants go through four stages (egg, larva, pupa, and adult) and live in colonies. Some species have more than one Queen. The workers have tiny ovaries. They can only lay male eggs because to create a female the egg must be fertilized, and it is only the Queen(s) that mate. In July and August you might see clouds of flying ants. These are the male drones and the Queens during their mating flights. Once the Queen has mated she will tear off her wings and find a suitable nesting site where she will lay thousands of eggs. The drones die after mating. This is a good time for birds, who eat the ants, so that only a few Queens survive to start new colonies. Queen ants have the longest lifespan of any insect, living up to thirty years.

Some of the worker ants are known as soldier ants. They have especially big heads which they use to plug the hole into the nest. When a worker ant returns it must touch the soldier ant's head to let it know that it belongs to the colony. Each colony has a distinctive chemical profile that enables members of the colony to recognise one another, but also alerts the colony of intruders. The Argentine ant has been found on every Continent except Antarctica. Scientists were surprised to discover that Argentine ant super-colonies in North America, Australia, Europe, and Japan all had the same chemical profile, making it the largest known super-colony in the world. It is thought that it contains 300 million ants.

Scout ants leave the nest in search of food. When they find a food source they eat some to check that it is edible, and if it is to their liking, they then return to their nest in a straight-line, laying a scent trail in order to help the other worker ants to find the food. Scout ants can observe and recall visual cues which help them navigate their way back to the nest. Once they get back, the workers go in search of the food, following and re-enforcing the scout's pheromone trail.

Some ants have symbiotic relationships with plants. They live in the plant's natural hollows and in return they defend the plant from herbivorous mammals and insects and even prune parasitic plants which attempt to grow on the host plant. Ants also help distribute the seeds of some plants, such as wild strawberries and wood violets. These plants seeds have a fleshy structure called eliasome which ants love to eat. Once the eliasome is eaten the ants discard the rest of the seed somewhere else, sometimes as far as 230 ft (70m) from home.

Ants sometimes herd other insects. They like eating the sugary secretions of insects such as leafhoppers and aphids. Leafhoppers sometimes leave their young with ants to raise in order that they can have a second brood. Some ants have been observed herding the aphids from plant to plant to keep an important food source close by. A single colony of ants can eat 200 litres of aphids' 'honeydew.' Ants are good at keeping the aphid population under control which is extremely helpful for gardeners. Aphids are all female. They lay unfertilized eggs which become larvae and then are blown by the wind onto nearby plants where they immediately start munching. They can inflict terrible damage to trees, especially beech trees, by leaving weeping wounds which them make the tree susceptible to pathogens and fungi. Normally when aphids want to move on they grow wings, but ants bite off the wings and exude chemicals from the aphids which make them slow down. John Whittaker at Lancaster University has noted that beech trees benefit from ants. Leaf loss in beeches not settled by ants is six times higher than those that have ant colonies living on or near them.

Some ant species enslave other ant species, forcing them to do all the chores around the colony.

Like bees, ants communicate in several different ways. They can convey different messages through body language and stroking and touching. They can also make different sounds, which are too low for our ears to hear, which have different meanings. Ants do not have ears, but they can detect and understand vibrations. They have compound eyes, like flies. Queens and drones have additional ocelli, like bees, but scientists do not understand why. They think it is something to do with taking in more light.

Ants do not have lungs, but they do have slits in their exoskeletons which allows oxygen to circulate around their body.

When I was about seven I went up Helvellyn in the Lake District for the first time with my dad. On the way down I needed a rest and sat on a grassy hillock. Unfortunately for me it was not a hillock but an ant hill. I got bitten all over my bottom by ants. Nearly forty years on I can still remember how painful it was. The reason it hurt so much is that the ants injected formic acid into me. It gets its name from the Latin name for ants, *formica*. The English naturalist John Ray was the first person to isolate formic acid in 1671. He distilled the crushed bodies of dead ants in order to extract the acid. It is one of the simplest organic acids and is used as a preservative and antibacterial agent in livestock food.

The ants do not just emit the acid as a defence mechanism, scientists at Martin Luther Halle- Wittenberg University in Germany have discovered that ants use their formic acid to disinfect themselves and to kill off harmful bacteria in their stomachs. Some species of ants douse themselves in formic acid to neutralise venom from other ants. Tawny crazy ants douse themselves to counteract the effects of fire ant venom. Researchers do not know how this works yet.

Butterflies and moths are also pollinators that we can observe in our gardens. The differences between moths and butterflies are not clear. Most moths come out at night and most butterflies can be seen during the day. Moths tend to be furrier than butterflies. Butterflies tend to be brightly coloured to enable them to communicate with other butterflies, whereas moths tend to be more camouflaged in order that they can rest on a tree during the day and not be detected.

About 10% of all living organisms are lepidoptera (butterflies and moths.) There are about 2500 species of moths and sixty species of butterflies in the United Kingdom. About a third of these have been observed in gardens. Sadly, butterflies have decreased by 76% and moths by about a third[80] in the last four decades, and several species have already become extinct. This is bad news not just for our plants, but for the many animals, bats, and birds that eat butterflies and moths. Blue tits are estimated to eat fifty billion moth caterpillars a year. Moths and butterflies are also a good sign of biodiversity. An area rich in butterflies and moths normally has lots of other invertebrates too. However, within this statistic there are winners and losers. The stout dart has declined by 81% whereas the Devon carpet moth and Jersey tiger moth have increased by 526% and 861% respectively. The Clifton nonpareil had not been seen for about fifty years but has expanded its range in the last 20 years, colonising Southern Britain, the Midlands, East Anglia, and Wales. It is well camouflaged but has a bright blue underwing which it uses to confuse would-be predators just long enough for it to escape.

Butterflies and moths have four life stages- egg, caterpillar, chrysalis, and butterfly or moth. A butterfly or moth lays about 100-300 eggs, which range between 1-3 mm and vary in colour and appearance according to the species. The eggs are attached to the underside of a leaf with special 'glue'. Very few of the eggs will survive because they are an important food supply for many birds, insects, and small mammals. Small funnel shaped openings called micropiles are at the top of each egg. The micropiles allow water and air to enter the eggs. The eggs are encased in a harder outer shell called chorion. The larva inside the egg is nourished by the yolk.

After about three to eight days, the larva (caterpillar) hatches from the egg. The caterpillar has one job- to eat. It needs a lot of energy to metamorphosize into a chrysalis and butterfly or moth. They start by eating the chorion around the egg because it is rich in protein.

---

80 39% in Southern Britain and 22% in Northern Britain

They must eat 27,000 times their body weight in the few weeks of their lifecycle, increasing their body mass by a thousand times or more.

Caterpillars have 4000 muscles, far more than the 650 which humans have. They move by contracting their muscles in their rear segments and pushing their blood into their front segments. As they move forwards their blood pressure changes. No other animal moves in this way. They have six legs and five pairs of false legs called prolegs which help them to climb and hold onto plant surfaces.

They have six eyelets called stemmata in a semi- circle on either side of their heads, but do not have good eyesight. They can probably only differentiate between dark and light. You might have noticed a caterpillar moving its head from side to side. This helps it judge depth and distance. They breathe through openings called spiracles along their body.

Some caterpillars and plants have co-evolved. Some caterpillars have evolved to be able to sequester toxins produced by the plant and use them to protect them from predators. For example, the six-spot burnet moth's caterpillars eat the leaves of birdsfoot trefoil and convert the toxins into hydrogen cyanide. They can also produce the cyanide themselves. This makes them taste unpleasant and can be fatal to predators if consumed in a large enough dose. The cyanide is also used when the adult moths mate. The female six-spot burnet moth releases plumes of hydrogen cyanide combined with sex pheromones to attract a mate, and male six-spot burnet moths can pass it to the female during mating.

This is just one example of mechanisms used by caterpillars to protect them from predators. Some are camouflaged, others are brightly coloured to warn predators of their toxicity (although in some cases this is fake) and others emit a smell to ward of predators. Be careful about touching hairy caterpillars because their hairs are actually venom glands. The venom can cause skin irritation, eye damage, and respiratory problems in humans. Some caterpillars can make themselves look bigger. Others have a symbiotic relationship with ants. The caterpillars of the lycaenid butterfly communicate with ants through vibrations and chemicals. The ants allow the caterpillars into their colonies and in return the caterpillars provide food for the ants.

Caterpillars produce silk from modified salivary glands near their mouths. Only silk from moth caterpillars is used by the textile industry. However, the silk is very useful when it comes to the next stage of metamorphosis – the chrysalis (for butterflies) or cocoon (for moths). Some use it to secure the chrysalis or cocoon to a leaf and then create a chrysalis or cocoon around them, whereas other species use the silk to stitch together leaves to form a chrysalis or cocoon. The caterpillar is triggered to start making a chrysalis or cocoon by hormone called ecdysone which tell it to stop eating and to start metamorphosing. Chrysalises vary in texture and colour depending on the species. Some are disguised as buds (orange tip butterflies), curled leaves (speckled wood) or even bird droppings (black hairstreak.) Some contain unpleasant tasting chemicals. Despite this chrysalises and cocoons are frequently targeted by parasitic wasps. In order to avoid them hairstreak butterfly chrysalises 'sing' to attract ants, who take the chrysalis into their colonies where they can safely transform into a butterfly.

Inside the chrysalis or cocoon a dramatic change occurs. The caterpillar's body produces digestive enzymes which enable it to digest itself from the inside out apart from some special cells called imaginal cells which are retained and transformed into the moths or butterfly's parts and organs, such as its wings and eyes. Some parts of the caterpillar's body, such as its mouth and legs, are retained. The process takes about a fortnight. During this period, the chrysalis or cocoon loses half its weight and consumes a lot of energy. A couple of days before the butterfly or moth emerges, the chrysalis or cocoon changes colour.

The first thing that the butterfly or moth must do when it emerges from the chrysalis or cocoon is to pump blood into its four wings. This can take a few hours during which they are very vulnerable. Butterfly and moth's wings are actually transparent. The order which they belong to, Lepidoptera, means 'scaly wing', and this is an accurate descriptions because their wings are covered in thousands of tiny scales which reflect the light of different colours. These scales appear as a powdery substance in moths. The wings are made from two layers of chitin, which is the same protein used to make insects exoskeletons, with capillaries in between.

Scientists have recently made some fascinating discoveries about butterflies' wings. For a long time, scientists could not work out how butterflies could fly because their wings are so thin. Recently a Swedish scientist called Professor Per Henningsson has discovered that butterflies flap their wings before they fly. In doing so they create an air pocket between the wings which creates more power. Even more remarkably scientists at Harvard and Columbia University have discovered that butterflies use their wings to regulate their temperature. Regulation of temperature is critical in insects. Butterflies can only fly if the air temperature is above 13°C. On cooler days they must bask in the sun or shiver to try and warm up. However, their wings can rapidly overheat in the sun. Whilst their visible wing colour helps to absorb and reflect the sunlight it does not help the wing to thermoregulate. Naomi Pierce and her colleagues at Harvard have discovered that their wings are made of two layers, tiny tube-shaped nanostructures, and the thicker layer of chitin, which help the butterfly to radiate excess heat from its wings. This also enables the living cells in the wings to determine the direction and intensity of the sun and counteract its effects with certain behaviours, for example, closing their wings or tilting their wings away from the sun. In a number of species of butterflies, males have 'wing hearts' on their wings which beat several times a second and facilitate the flow of haemolymph (blood) through their scent pads, thus producing and emitting pheromones which attract the females. This was an important discovery because until this time scientists thought that the butterflies and moths only used haemolymph to emerge from the chrysalis and push their wings open for the first time.

Another early task for a butterfly is to assemble its mouthparts, which are in two pieces when it emerges from the chrysalis. It joins them together to form a proboscis, which acts like a straw. Butterflies can only eat liquids, normally nectar. They can unfurl the proboscis when they need it to suck nectar from a flower then curl it back under their chin when they are not feeding. Male butterflies drink from puddles. They incorporate minerals from the puddle with their sperm, thus making the eggs more viable.

Butterflies taste with their feet. They drum on the plant's leaves to release the juices and then spines on the back of their legs have chemoreceptors which detect if it is the right match of plant chemicals. In doing this the female butterfly can make sure she lays her eggs on the right plant and ensure that the caterpillar will have something to eat when it hatches.

Butterflies have photoreceptors (light-detecting cells) in their eyes which convert images into electrical signals which are sent to the brain. They have compound eyes, like flies, made up of thousands of tiny lenses. This enables the butterfly to see in all directions. However, they can only see within about a 10–12-foot (3-4m) range and anything beyond that is blurred. They can see ultraviolet light. There are ultraviolet markings on butterflies' wings which help them identify other butterflies. Plants also have ultraviolet markings which act like landing strips guiding the pollinator to the nectar or pollen.

Moths are often seen as the butterflies' poor relations. The butterflies are the superstars, and the moths are the Z-List celebrities. However, this is unfair. Ninety per cent of lepidoptera are moths. Like butterflies, moths have amazing abilities. Clearwing moths, bee hawk-moths, and the lunar hornet moths can mimic bees and wasps in order to avoid predators. Goat wasps are the heaviest moth in Britain and get their name because they smell like a goat (no -one seems to know why) Elephant moths are pink to attract a mate, and no two garden tiger moths are the same (even the wings of an individual can be different). The death's-head hawk-moth makes a noise like a Queen bee in order to sneak into hives and steal honey.

The peppered moth is a mottled white moth but in 1848 a black one was found in Manchester. By 1900 98% of the peppered moths found in cities were black. They are nocturnal and sleep on walls and trees during the day, so it was thought that they had evolved in order to camouflage with their surroundings, or that chemicals in the smoke darkened the moths or triggered a chemical reaction in them. Dr Ilik Saccheri at the University of Liverpool has recently discovered that peppered moths changed colour because of a mutation which dates back to about 1819, near the start of the Industrial Revolution. However, the mutation creates as many questions as it answers because Dr Saccheri discovered that it inserted itself into a gene called cortex, which is not known to play a role in pigmentation. Since the Clean Air Acts of the 1960s, the original mottled white coloured peppered moths have had a resurgence so it will be interesting to see what happens to the black ones.

Male moths can smell pheromones released by female moths by using highly sensitive receptors on their antennae. Male silk moths will follow a pheromone trail for miles, but the record goes to a male promethea moth which flew for 23 miles (37km) following a pheromone trail. He must have been extremely disappointed when he found a scientist, rather than a female moth, at the end of it.

Some moths live for such a short period of time that they do not need mouths.

Not all moths are nocturnal. There are more than a hundred species of diurnal moths in the United Kingdom. However, many people have probably noticed moths flapping around streetlights. No-one knows for certain why moths are attracted to the light, but it would seem that nocturnal moths might have evolved to use the moon in order to navigate, although other scientists dispute this because of the way in which the moths approach the light source, a circuitous route rather than keeping themselves at a constant angle to the light. Others think that moths dip when they see an artificial light because they have a dorsal light reaction. Like most flying animals they fly by keeping the lighter sky above them. When they see the artificial light it confuses them. However, even then it is not straightforward. Male moths seem to be more attracted to lights that female moths. Scientists in Germany have discovered that moths that live in areas with higher levels of light pollution are actually about a third less likely to be attracted to the artificial lights compared to moths that live in darker areas. However, the lights do have some effect on the urban moths, causing them to fly less, thereby effecting pollination and providing less food for other animals.

Like birds, some moths and butterflies are migratory. Painted lady butterflies, named after the 17th century fashion for ladies to wear a lot of make-up, are long distance migrants making an incredible nine thousand-mile (14480 km) round trip from North and West Africa to Northern Europe (as far North as Iceland and Northern Norway) in the Spring and then back to Africa in the Autumn. They also have closely related cousins who migrate across North America and across Australia. They fly at high altitudes of more than 1640 feet (500m) at speeds of 30 mph, incredible for a butterfly which weighs less than a gram and has a brain a sixth of the size of a pinhead. They cover about a hundred miles

(161km) per day, and as butterflies have such a short lifespan it can take six generations to successfully complete the entire journey. They can navigate a journey that neither them, their parents, or grandparents have seen before by using a solar compass in their antennae. It seems that they can adjust their migratory patterns according to local weather conditions and topography. New generations know to fly South in the Autumn and North in the Spring. They know when to migrate by the lengthening or shortening of the days. As part of their journey about one million come to Britain every year but every ten years there is a population boom and millions come. It happened a few years ago and my garden was full of them. They are beautiful butterflies with just over a two-inch (6cm) wingspan so easy to spot.

Another type of lady you can see in your garden is a ladybird. There are 47 species of the ladybird in the United Kingdom and more than five thousand worldwide. It is not known why they are named after the Virgin Mary, but it is something found in most European languages. In French they are *bete de la vierge,* in German they are *marienkafers* and in Spanish they are *mariquitas.* One theory of how they became associated with Jesus's mother is that in early images of the Virgin Mary she always wore a red cloak. Another theory is that the seven spots on the most common ladybird in the United Kingdom, called unsurprisingly the seven-spot ladybird, represent the seven sorrows of the Virgin Mary – Simeon's prophecy, the flight to Egypt, the three days that Jesus was lost in the Temple, the meeting of Jesus and Mary on the way to the cross, the crucifixion, Jesus being taken down from the cross, and the burial of Jesus. I am not sure that either explanation is particularly convincing because ladybirds come in a variety of colours and have a variety of spots, so I suspect the real reason is lost in time. Another story which seems to originate in the Middle Ages is a bit more plausible. It is said that there were a plague of insects destroying crops in Europe, causing people to starve to death. People prayed to the Virgin Mary, and she sent ladybirds to kill the pests, and so the coloured beetles were called ladybirds in her honour. Whatever the origin of their name, possibly because of their connection with the Virgin Mary, they have long been associated with good fortune and it is thought that if you kill one you will be dead by the next day.

The Wonders of the Wild Places

The first written record of it being called a ladybird in this country is in 1674, although Shakespeare used is as a term of endearment. The nurse in Romeo and Juliet says, "...*what lam, what ladie bird.*" Prior to this time the ladybird seems to have had lots of regional names. In Norfolk it is known as Bishy Barney Bee and is associated with the following rhyme-

Bishop, Bishop Barnabee
Tell me when my wedding will be
If it be tomorrow day
Take your wings and fly away
Fly to the east, fly to the west,
Fly to them that I love best.

The origins of this are unknown. Some argue that it derives from Bishop Bonner who persecuted protestants during the reign of Mary I. He might have worn a red cloak. Others believe it might relate to St Barnabas whose feast day is the 11th of June when ladybirds are commonly seen.

As a child I would say the rhyme *Ladybird, ladybird, fly away home, your house is one fire, and your children are gone*. It is only as an adult that it occurs to me what a strange rhyme it is. Like many nursery rhymes its origins are lost in time, but there are many theories. Some link it with the Great Fire of London in 1666, others think it was a code used by Catholics during Tudor times when they were being persecuted. Another story is that farmers sang it before they burnt their crops to give the ladybirds chance to escape. Another possibility is that it was a way of teaching children to look after them and avoid bringing bad luck to the household.

The rhyme is first recorded in a written form in 1744 in *Tommy Thumb's Pretty Songbook*. It is a different version to the one I used to recite.

Ladybird, Ladybird, fly away home,
Your house is on fire and your children are gone,
All except one, and her name is Ann,
And she hid under the baking pan.

The Wonders of the Wild Places

The name and gender of the child and the kitchen utensil appears to differ depending on which part of the country you live. For example, in Peterborough the child is John, and he hides under a gridle stone. Versions of the rhyme are sung in several other European countries and across the United States of America.

Ladybirds are the gardeners' friends because they are natural pest controllers. They lay hundreds of eggs in the colonies of aphids and other plant-eating pests. As soon as the ladybird larvae hatch, they begin munching their way through the aphids or other pests. Within its three-to-six-month life each larva can eat up to five thousand aphids.

Like all insects they have a four-part life cycle from egg to larva to pupa and then to adult ladybird. When the ladybird emerges from the pupa stage they are bright yellow. Over the next few hours their wing case hardens, and their colour and pattern develop.

If attacked they emit a yellow, unpleasant tasting fluid. I remember it being on my hands as a child because I was always picking up ladybirds. This substance is reflex blood and is emitted from their leg joints. It is rich in toxic alkaloids. Most birds and animals associate them with an unpleasant taste, but they are eaten by swallows and swifts, spiders and beetles and the harlequin beetle (more of which are the end of the chapter.)

Another 'insect' which appears in nursery rhymes are spiders. Actually, spiders are not insects, they are arachnids. They are also arthropods because they have eight legs. They have a different anatomy to insects and move differently.

There are about 49,623 species of spiders 5 in the world ranging from the enormous Goliath tarantula that eats birds to the tiny 0.03cm long Patu digua in Samoa (although like many things in nature there are other candidates for the largest and smallest spider title).

One of the most amazing structures in nature is a spider's web. Not all spiders make webs. Out of the 37 species of spiders in Britain only seventeen weave webs. There are different types of webs. They vary a lot in fragility from the robust funnel webs of house spiders which can last several generations of spiders to the very fragile orb webs which are prone to get destroyed by the wind and rain (and their stickiness can be rendered ineffectual by pollen and dust) so they have to be made daily.

However, the webs are not wasted, the orb spider ingests the old web and recycles the ammino acids. It can reuse them half an hour later.

Spiders construct their webs out of silk. It stores silk protein in liquid form in its gland store. The spider guides the silk out its spinnerets on its abdomen. As the liquid ammino acids passes through a canal it is acidified, which changes its structure. It contains crystalline proteins which makes it strong and amorphous proteins which give the silk its elasticity. The motion of pulling the silk from the spinnerets helps it solidify. It takes about an hour to produce a web.

Spiders actually make seven different types of silk from seven different glands. Each type of silk has a different purpose and properties. Ampullate major is the dragline. This is used for the outer rim and spokes of the web. It is also the silk you see when a spider suddenly lowers itself down in front of you. It also uses the dragline to find its way home. Ampullate minor is used as temporary scaffolding whilst the web is being constructed. Flagelliform is capture spiral silk. The unfortunate prey, usually insects, are caught in the sticky silk. They also have tubuliform silk which they use to create egg sacs; aggregate, which is the glue which holds the web together, and the piriform gland which anchors a thread to another thread.

Spider silk is one of the most wonderful things in nature. It is stronger than steel but weighs a sixth the density of steel. It is thirty times smaller than a human hair. It is also very flexible and elastic. It can stretch five times its length without snapping. It is exceptionally durable and can withstand temperatures of -40°C. It is also antibacterial. Maybe for this reason the Ancient Greeks used to make bandages out of spider silk. Scientists at the University of Wyoming are currently considering whether it could be a biodegradable alternative to Kevlar, which is used, amongst other things, to make bullet proof vests. The problem at the moment is how it could be mass produced. As it is unlikely that we will see spider farms, scientists are trying to make a synthetic version.

Young spiders, called spiderlings, can use silk threads to 'fly.' They cast the silk thread into the wind and then allow themselves to be carried, sometimes miles, by the wind. This is called ballooning. Erica Morley at the University of Bristol has discovered that spiders use static electricity to balloon up to three miles (5km) above the earth. They can travel for thousands of miles using this technique.

Spiders normally has eight eyes (although they can have fewer, although it is always an even number) but they do not have good eyesight. They detect prey by sensing vibrations. They can detect if something as small as a thousandth of a human hair has entered the web. They then side up to the victim as it tries to free itself and use setae in their feet to wind sticky silk (acinform) around the struggling insect. They inject their victim with poison which contains enzymes which liquidise the prey and then the spider sucks up the gloopy remains. One spider can eat hundreds of small flies a day.

Many species of spider are cannibalistic. It is normally the female who eats the male, particularly if he has not bought her a gift or she is not impressed with his courtship. It is a tough life being a male spider.

It is not just insects and arachnids that you can see in your garden, sometimes you can see a mammal. When I was a child the mammal I would see most in my garden was a hedgehog. We had one living in the compost heap, and one living in the hedge over the road, and at least three living in the school playground. I saw at least one hedgehog almost daily during my childhood. Sadly, that has all changed now. Just in the last decade hedgehogs have declined in urban areas by about a third and between 30-75% in rural areas, with the most dramatic decline in the East of England. There is a real risk of these iconic animals which have been around for 15 million years becoming extinct in Britain in the next decade or two.

Hedgehogs are Britain's only spiny mammals. The spines are made from keratin, which is also what human hair and nails are made of. They are born without prickles but develop them after a couple of days. An adult hedgehog has between 5000-7000 spines.

They have poor eyesight and rely on their sense of smell to hunt invertebrates. If you are lucky enough to have them in your garden do not give them milk because they are lactose intolerant. However, they can eat cat or dog food.

One of the reasons they have declined is because of loss of habitat. They roam about one or two miles (2-3km) during a night, over ten to twenty hectares. It helps them if they there are spaces under fences that they can squeeze under.

They hibernate from late Autumn until the Spring. Their body temperature drops, and they fall into a deep torpor, saving energy. If you find a hedgehog do not disturb it. However, it is normally a bad sign if you see one during the day so try and catch it with a towel and take it to a vet or animal hospital. Many hedgehogs become malnourished.

Parks

If you do not have a garden maybe you are lucky enough to live near a park. I live opposite one, and it is a huge asset to the town where I live, as well as being a great place to spot nature.

There are approximately 27,000 public parks in the United Kingdom, 300 of which are registered as being of national importance. More than 37 million people (57% of the U.K. population) regularly uses parks. If the area of the top hundred parks in the United Kingdom was added together it would cover an area larger than the Isle of Wight.

One of the first public parks was Birkenhead Park, created by landscape architect, Frederick Laws Olmstead in 1847, and said to be the inspiration for Central Park in New York. The first park created in Britain was Miller Park in Preston which was created by out of work cotton workers in 1833. There were numerous Acts of Parliament in the Victorian period, which encouraged wealthy industrialists to donate land and money to create parks for their workers. David Tibbotts of Greenspace explains that the parks were created to, "...please the cultural eye and were meant to be a glorification of God's nature. People were meant to be awestruck by... the refinement of the planting, the heightened landscapes, and the finest horticultural standards."

Many of these parks were designed by the renowned gardener, architect, and landscape designer, Joseph Paxton. He was born in Bedfordshire in 1803. He was the seventh son of a farmer. In 1826 he was appointed Head Gardener at Chatsworth in Derbyshire. Paxton stayed in that position until the 6th Duke of Devonshire, William Cavendish, died in 1858. With the Duke's encouragement he transformed the gardens at Chatsworth, and designed glasshouses, and planted an arboretum. During this time, he also began to accept commissions to design other gardens and parks, including Prince's Park in Liverpool, People's Park in Halifax, and the Spa gardens in Scarborough. He also found time to design one of the first municipal cemeteries in Britain, London Road Cemetery in Coventry, as well as designing Crystal Palace, and being the Liberal Member of Parliament for Coventry from 1854 until his death in 1865. He is credited as cultivating the Cavendish banana, the most consumed banana in the Western world.

Paxton was also fascinated with collecting exotic plants from far flung places across the globe, although he sent other people on the expeditions. One expedition ended in disaster when two of Chatsworth's gardeners drowned in California.

Plant- collecting was popular in Georgian and Victorian times. People scoured the Continents, especially the Tropics, looking for exotic plants and trees. Plant collectors and seed hunters shipped back plants, seeds, and saplings to Britain, where they were in great demand for country estates and parks. Unfortunately, this had a devastating effect on our natural world, which we are still experiencing the repercussions of, because in some areas all the plants were taken.

Britain's first official roving plant collector was Francis Masson, a Scottish gardener, who worked with Joseph Banks (who famously went on expeditions with Captain Cook) at Kew Gardens. In 1772 Banks send Masson to South Africa and Masson spent two years recording the plants that he saw, including gladioli, irises, and heathers. Unlike other plant collectors Masson painted the plants he found, rather than removing them from their natural habitat. He liked South Africa so much that he settled in Cape Town and created a wonderful garden.

A good example of just what plant-collectors went through to find and collect 'new' species is a man who spent half his life about ten miles from where I currently live. Richard Spruce was born at Ganthorpe, near Castle Howard in North Yorkshire, in 1817. He was a sickly child so was encouraged to spend time outside, where he developed a love of botany, especially bryophytes (mosses and liverworts) which he described as, "... the underdogs of the plant world." He became a teacher, like his dad, but in 1848 he was advised to go abroad because of his health. He went to the Pyrenees and spent a year collecting and recording species. During his time there he discovered seventeen species new to science and increased the known bryophytes in the region by more than 300 species. Due to his success in the Pyrenees he came to the attention of Sir Joseph Hooker and George Bentham at Kew, who asked him to go on an expedition to South America. Despite being in poor health with a range of real and imagined conditions he spent fifteen years travelling throughout Ecuador, Peru, Bolivia, Brazil, and Venezuela, exploring the Andes and Amazon looking for new species. He worked in terrible conditions and had to battle with parasitic insects and bouts of malaria. The latter is spread by infected mosquito bites and is a disease which still kills millions annually across the world. At the time Spruce was in South America Britain and other European nations were colonising large parts of South America, Africa and India, areas where malaria was, and still is, endemic. Therefore, it became critical to the government to have a regular supply of quinine, the drug used to treat malaria, which is derived from the bark of the chinchona tree (it was named by two French doctors in 1820 after the Amerindan name for the tree, *quinaquina* meaning bark of barks.) By the 1850s two million people a year were dying of malaria in India and millions more were seriously ill with it so, in conjunction with the East India Company, the government decided something needed to be done about the situation. Another Yorkshire explorer, Clements Robert Markham, came up with a plan to smuggle seeds and plants from South America to India in order that chinchona trees could be grown in India, thereby overcoming the problem of transporting quinine from South America, and ensuring that the British had their own supply of quinine. Markham knew that Spruce was already in South America so recruited him to help gather seeds and plants and send them to India. Working in horrendous conditions, and despite having to deal with illness, semi-paralysis, and avoiding warring factions and civil wars, Richard Spruce somehow managed to smuggle more than 100,000 seeds of the chinchona tree and 600 plants to India. In doing so he helped save a lot of lives, but it broke him health wise and financially. He was forced to eventually return to the United Kingdom, back to where he grew up, near Castle Howard. He spent the last 27 years of his life paralysed from the waist down, sorting through the plants he had brought back. Two hundred species are named after him, although not spruce trees which get their name from an old word for Prussia. He was also fascinated in Amazonian languages and customs and brought back the vocabularies of 21 Amazonian languages and details of their customs and traditions, as well as detailed maps of three previously unexplored rivers. I think Richard Spruce would be proud to know that the University of York has had a major breakthrough in the treatment of malaria. Artemisin is a compound extracted from a plant called sweet wormwood, or *Artemisia annua*. Sweet wormwood has been used in Chinese medicine for at least two millennia. The University of York has created a hybrid version of the plant which has a higher concentration of artemisin in its leaves. The new hybrid is now being grown in China, and is being made into the anti-malaria drug, artesunate, which can effectively cure malaria. According to the Lancet, as a result of this development, malaria could be eradicated by 2050. Incidentally this is also an important development for another reason. There is strong evidence to show that artemisin can effectively kill cancer cells. Cancer cells needs iron to be able to spread. However, when the iron and artemisin enter the cancer cell together they combine to form atoms called free radicals which kill cancer cells without harming normal cells. Only small studies have taken place so far but artemisin has already been shown to successfully treat and eradicate aggressive cancers, such as colon cancer and melanoma, by slowing the spread of tumours, causing cancer cells to self-destruct, and stop division of cancer cells and stop them from spreading. Larger studies are required but it looks likely that this compound from the sweet wormwood plant might play a big role in medical advances and might save millions of lives in the years to come.

    Originally it was hard to transport live plants back from North or South America, or the Tropics or Africa, back to the United Kingdom and plants were often dead by the time they got here. However, Nathaniel Ward made it much easier to transport plants and trees by inventing a glass box, known as the Wardian box, which Victorian plant and seed collectors used to transport their valuable cargo to the United Kingdom. The box worked like a greenhouse, allowing sunlight to the leaves, resulting in condensation which dripped to the roots, allowing the plant to stay hydrated on its long journey. After the glass tax was repealed in the mid-

19th century these Wardian boxes became much more affordable, leading to a huge increase in plant and seed collectors. This sparked the Victorian craze of pteridomania or fern collecting. Professional fern traders scoured the globe, clearing entire areas of ferns. The largest fern, the Royal fern, was collected by so many people that it now only exists in the wild on the Isle of Man. As real ferns became scarcer the craze changed to using ferns as a design for wallpaper, furniture, and dresses.

Orchidelirium, the craze for orchids, which also took place in the mid-19th century did even more damage to the natural world. The owners of hothouses (early greenhouses) sent their gardeners to the Tropics to collect orchids. They cut down trees to reach the orchids living in their canopies and deliberately eradicated orchids in one area to stop others from getting them and to make their specimens rarer and therefore worth more. Most of the orchids did not survive for long if they made it back to the United Kingdom because no-one knew how to look after them. Many died in unsuitable hothouses.

Parks were an opportunity for middle- and working-class people to see non-native plants and trees. One of the things they would have seen are sequoias, which are very tall, long-living trees which are native to North America. In 1852 Augustus T Dowd became the first white man to see them growing in Yosemite Valley. From that moment people became fascinated by their size. In 1854 an ex-miner called George Gale cut up an old sequoia and sent it to New York where it was exhibited as one of the wonders of the world. People thought it was a hoax because they could not comprehend its size. A year later an 'arboreal amusement park' was developed at Calaveras Grove in California and it soon became a popular tourist attraction. People could dance in one of the sequoias, and another one was used as a bowling alley.

We are probably most familiar with one type of sequoia, the giant redwood, the largest trees in the world. They can grow more 328 ft (100m) high. It is native to California and Oregon. It has a fireproof bark. The soft bark traps air, insulating it and enabling it to withstand moving flames. However only older redwoods have this ability. The wood is used to make telegraph poles, coffins, and organ pipes.

Another tree brought back from the Americas was the monkey puzzle tree, also known as the Chilean Pine. It is native to Chile and Argentina. It arrived in Britain by way of a plant collector called Alexander Menzies who was on a voyage surveying the Pacific. He and other officers on the ship were invited to a dinner given by the Viceroy of Chile. Menzies noticed some strange nuts served with dessert and secreted them back to the ship where he planted them and managed to grow five saplings which he successfully got back to England. When he arrived home he gave them to Joseph Banks. The story surrounding its vernacular name originates in 1834 when a young Chilean pine was planted in Cornwall and supposedly attracted a group of bystanders, one of whom remarked that the unusual branch structure of the pine would, "...be a puzzle to a monkey." When a wealthy Victorian lady, Marianne North, travelled to Chile in 1884 to paint a Chilean pine she wrote that it was known as a puzzle-monkey tree in England, "...rather unreasonably because there are no monkeys in Chile to puzzle."

Churchyards

There are about 10,000 churchyards in the United Kingdom. Churchyards and cemeteries are great places to see nature. A survey in 2019 found that they were particularly good places to see rare species. One hundred and forty protected species were found in churchyards and cemeteries, including 70 protected plants and 49 protected animals. Highgate Cemetery in London covers 36 acres and has a high level of biodiversity. Some of their mausoleums are home to the exceedingly rare orb-weaver spider, a cave dwelling spider, which was only discovered nine years ago, although scientists suspect that they have probably being living in the mausoleums for at least a century.

Bats are a protected species you might see in churchyards and cemeteries. There are eighteen species of bats in the United Kingdom, but worldwide there are 1400 species which accounts for a fifth of mammals worldwide (bats represent a quarter of all mammals in the United Kingdom). They are the only mammals in the United Kingdom which can fly. However, their wings are really hands, which have adapted for flight. The family name for bats is chiroptera, which means *hand wing*.

The Wonders of the Wild Places

They are much more flexible and adaptable than birds. Bats can hover and fly backwards really quickly. You might have seen a bat hanging upside down when it is asleep. It can do this because it has a locking mechanism in its toes and claws which stop it from falling. Bats sleep in this position, because unlike birds their legs cannot support their bodyweight, so they are unable to launch themselves from the ground.

Bats are good indicators of the state of biodiversity in the area. Over 500 species of plants rely on bats to pollinate them and through excreting seeds they help create new woodland. As some British bats fly up to nineteen miles (30 km) to feed, they can transform a large area.

Bats are the top predators of nocturnal insects. They must eat half their own bodyweight of insects per night to survive. The smallest bat in Britain, the pipistrelle (from the Italian word meaning *little piper)*, can eat up to three-thousand insects a night. This helps farmers and gardeners because they do not need to use pesticides. Bats locate their prey in the same way that whales do, echolocation. You have probably heard them making high pitched noises as they fly. A bat contracts its larynx two-thousand times per second during echolocation. This is the fastest muscle in mammals, a hundred times faster than the muscles we use to blink. The largest[81] and loudest bat in Britain is the noctule bat whose call is so loud it must close its ears when it calls so it does not deafen itself. The sound waves provide the bat with information about the size and shape of the insect, and which direction it is travelling in. Using echolocation bats can locate prey within 82-98 feet (25-30m).

Moths have several ways of detecting and confusing bats. The reason that moths' bodies are furry is so they can deflect soundwaves in different directions, to confuse the bat. If moths fall to the ground, it also makes it harder for the bat to find them. Bats can hear at much higher frequencies than humans, 212 kilohertz compared to the 20 kilohertz that humans can hear. Most moths cannot hear at these high frequencies, but Hannah Moir at Leeds University has discovered that a moth called the greater wax moth can hear the highest kilohertz in the animal kingdom, an incredible

---

81 The Nodule bat has a wingspan of 13-18 inches (33-45 cm), much smaller than the largest bat in the world which lives in Java and has a wingspan of 5ft (1.5m).

300 kHz. Hannah Moir believes that they can hear at these frequencies in order that they can hear bats, but also so that they can distinguish between the sound of a bat and another greater wax moth, making reproduction easier. What makes the greater wax moth's ability to hear such high frequencies even more remarkable is because they have extremely simple ears with just four receptor cells, far fewer than the 20,000 receptors in a human ear. Astonishingly, the greater tiger moth has evolved the ability to interfere with the bat's echolocation by mimicking the sound of the bat, confusing the bat long enough so that the moth can make its escape.

The phrase *blind as a bat* is actually incorrect because bats have good eyesight. However, their hearing is exceptional. Incredibly, a brown-eared bat can hear a ladybird walking along a leaf.

Bats hibernate in the winter. The best time to see them is on a summer evening when you might see them flying over a garden or field, or if they are a Daubenton's bat, over water.

A tree which grows in many churchyards is the oldest species of tree in Britain, the yew tree. The oldest yew tree in Britain, and in fact Europe, is reputed to be the Fortingall Yew[82] which grows in a remote churchyard in Stirlingshire. I have seen it, and although it looked old and impressive, it was amazing to think that it is believed to be about 5000 years old. There is a legend that Pontius Pilate, the man who was reputedly responsible for Jesus's death, was born under this tree. It is alleged that his father was a diplomat on an expedition to see a local Pictish King when Pontius Pilate was born. Of course, there is no way of proving whether this is true or exactly how old the Fortingall Yew is. Normally trees are quite easy to date. Dendrochronological core samples can be taken which give reasonably accurate indications of the tree's age, or alternatively measure the circumference of the tree in centimetres about a three feet (one metre) off the ground and divide it by 2.5 to get a fairly accurate estimate of its age. However it is much harder to date a yew tree because they tend to become hollow as they get older, and because they have the ability to regenerate, which is one of the reasons they are so often seen in churchyards, and

---

[82] The yew tree at St Cynog in Defynog, Wales, is a similar age and might be older than the Fortingall Yew.

younger versions of the yew can merge with the original yew.[83] However, it is believed that there are ten yew trees in England which predate the 10th Century and that at least 500 churchyards in England have yew trees older than the churches they are next to. Britain is home to more ancient yews than any country in Europe. The Ancient Yew Group states there are 978 ancient or veteran yews (over 500 years old) in England and 407 in Wales, whereas France has 77 and Germany and Spain only have four each. The reason that Europe has so few ancient yews left is because of the huge demand for longbows, which were made from yew (Spanish and Italian yews were particularly sought after), in the 14th and 15th centuries, which led to the destruction of European yew forests.

There is a lot of debate about why yews so often grow in churchyards. Yew is associated with death and resurrection. The famous 18th century naturalist, Gilbert White, considered the various theories regarding why yews tend to appear in churchyards- shelter, to keep livestock out, associations with immortality- but could not find any evidence. In the 1940s Vaughan Cornish did extensive research about yews in churchyards. He believed that yew trees were important to early British people because of their association with immortality. He thought that the early Christian church had adopted this association. He suggested that it might have been the Normans who started the tradition of building churches near yew trees. He based this belief on discovering that most ancient yews in England and Wales appear in the churchyards of Norman churches. The yew trees were on the South side of the churches, which was the side of the church used in funeral possessions so tied in with the associations with death and immortality. However, he did not find any evidence that the trees predated the churches, although certainly in the case of Fortingall and St Cynog they must do.

After having dreams in which he was instructed to look for the Tree of the Cross, Allen Meredith visited most of the ancient yews in England in the 1970s and measured them. After doing some calculations he believed that 500 yews were older than a thousand years old, which might tally

---

83 At one time there was a belief that the Fortingall Yew was more than one tree, but DNA analysis has confirmed it is one tree.

with Vaughan Cornish's belief that most of them date back to the Norman conquest.

However, bearing in mind that most of the yews are older than the churches they are next to, it cannot be said definitively whether architects designed the church around the yews or whether the yews already signified a significant Neolithic spiritual site.

All the parts of a yew are extremely poisonous but the anti-cancer drug, Taxol, has been developed from the bark of the Pacific yew, and has been used in the treatment of ovarian, prostate, and breast cancer since the 1990s.

Urban Streets

Although foxes are in decline in rural areas, possibly because their populations are still recovering from centuries of hunting, they are a species which is thriving in urban areas. Foxes are very adaptable omnivores. The red fox, which is the species we have in the United Kingdom, is found in every country in the Northern Hemisphere and Australia. There are an estimated 430,000 foxes in the United Kingdom, one for every 150 urban dwellers. The urban fox population increased from 33,000 in 1995 to 150,000 in 2017. However, there was a 42% decline in urban foxes in 2018. Scientists at the University of Bristol are still trying to discover why this happened. It is known that foxes regulate their own numbers by varying the number of cubs they have each year. Despite this although other countries have urban foxes no other country has as many urban foxes as Great Britain. It is thought that they live in pretty much every town and city in the United Kingdom. A study by Brighton and Reading universities has discovered that Bournemouth has the largest concentration of urban foxes, with 23 per square kilometre, followed by London, Bristol, and Newcastle-Upon-Tyne. Foxes have lived in London since at least Victorian times, but many more moved into the suburbs during the Inter-War years. When I lived in Leeds I saw more foxes than I have done in the almost forty years I have lived in the countryside. One fox was so tame that it would come and peer through the patio doors, hoping we would put some food out for it.

Foxes live in social groups although they forage for food on their own. In the country they live in family groups with the parents and cubs, but in urban areas there might be several adults in the group. In rural areas their territories cover about nineteen square miles (49 km²), but in urban areas, where food is more abundant, its normally about 4.5 square miles (12km²). Foxes have scent glands in their feet which create scented trails which enable them to find their way around at night.

They have fantastic sense of vision and hearing. They also have a complex language which incorporates 28 different calls. They can run up to 30 mph. They do not chew their food. They just use their sharp teeth to cut the food into manageable pieces that they then eat whole. Rural red foxes eat about 95% meat, but in urban areas, because of the large amount of household waste, only half their diet consists of meat. They have fantastic immune systems which enable them to eat rotten food. Urban foxes are also useful pest controllers and have been shown to keep the rodent population under control.

Scientists have discovered that foxes use the earth's magnetic fields to hunt. Jaroslav Červený spent two years studying red foxes in the Czech Republic and realised that when the foxes were mousing (which is when they jump on their prey) they mainly jumped in a north-easterly direction regardless of time of day, season, wind direction, or weather. He thinks that the fox is using the earth's magnetic field as a 'rangefinder' to help it estimate the distance between itself and the prey. In the Northern Hemisphere the earth's magnetic field tilts downwards by about 65 degrees below the horizon. Červený thinks that as the fox creeps up towards its prey it waits until it can hear it and then uses the magnetic field to work out the distance. When they jump from a north-easterly direction, they are successful about three-quarters of the time, as opposed to 60% when they jumped from the south-west, and about a fifth of times when they jumped from other directions. If Červený's research is correct then the fox is the first animal known to use the earth's magnetic field to hunt, and the first to use it to work out direction. The fox senses the earth's magnetic field by using a protein called cryptochrome in its eyes. Cows and deer have also been shown to use the earth's magnetic field. They align themselves in a north-south alignment regardless of conditions, but another Czech scientist, Hynck Burda, has discovered that the electric magnetic field can be disrupted by high voltage powerlines, and this can confuse deer and cows.

Incidentally, recent research has also discovered that humans also use the earth's electromagnetic field. Scientists at the California Institute of Technology found that people form a distinctive brain wave pattern when exposed to the earth's magnetic field. It only appears when the magnetic field points North, which is to be expected because all the people in the study lived in the Northern Hemisphere, where the magnetic field is in a north-east direction. Scientists also discovered that the human brains contained magnetite, which has been detected in the beaks of pigeons and other birds.

Foxes appear in folklore in many cultures, especially European and East Asian countries. They are associated with cunning, trickery, and transformation. Aesop was the first person to write about them in 4BC. In the Middle Ages there was a popular story about Reynard the fox. Geoffrey Chaucer used this story as the basis of his *Nun's Priests' Tale* in *Canterbury Tales*. In Victorian times Joel Chandler Harris wrote about *Brier Fox* and Beatrix Potter includes a wily fox in the tale of *Jemima Puddleduck*. More recently authors such as Roald Dahl in *Fantastic Mr Fox* and Ted Hughes in the *Thought Fox* have turned to foxes for inspiration.

A lot of words and expressions are associated with foxes, whether it is to outfox someone or to look foxy. The word *shenanigans* meaning a deceitful trick or mischievous behaviour comes from the Irish Gaelic word *siannachuighim* which means *to play the fox*.

A recent study by Dr Kevin Parsons at the University of Glasgow's Institute of Biodiversity, Animal Health, and Comparative Medicine discovered that urban red foxes in London have evolved differently to their rural cousins in the surrounding countryside. The urban foxes have smaller brains and different snouts to help them forage for food in an urban environment.

Foxes are not the only mammals who have evolved or adapted to suit their urban environments. A study in the U.S.A. showed that urban cotton mice and meadow voles have bigger brains than their country cousins, which enables them to respond efficiently to an urban environment.

Urban grey squirrels use visual cues rather than audio cues and are more territorial and reactive than their rural counterparts.

When Victorian planners built a lot of our suburbs they incorporated trees into their designs, and many of us live in tree lined streets. Urban trees do not just look nice, they also have significant effects on the environment and on human health. They help reduce temperatures, flooding, and pollution; improve physical and mental health, influence the birth weights of babies born in areas with a large canopy of trees, and help to reduce crime. However, it has been shown that mature trees are 92% better than saplings at helping the urban environment. Currently sixty street trees are cut down every day which is a concern. However, urban councils are beginning to understand the importance of urban tree coverage. In Bristol they have a tree replacement scheme, ensuring that mature trees are only felled when it is unavoidable, and that when they have to be cut down, because of disease, trees which have been shown to be particularly effective at combating environmental problems, such as silver birch, yew, and plane trees, are planted in their place.

London plane trees do not originate in London although they are prevalent in the capital. They are a hybrid of the American sycamore and the Oriental plane. No-one knows how the hybrid came about. One theory is that they hybridised in Spain before somehow coming to Britain in the 17th century. Another possibly more plausible theory is that the hybrid was created by a 17th century botanist, plant collector, and gardener called John Tradescant in his garden in Vauxhall. However they got here London planes are useful trees in cities because it catches pollution in its bark and hairy leaves. It has a flaky bark which breaks away in big pieces, trapping the pollution. A 2011 study found that London planes in Greater London removed 850 – 2000 tonnes of PM10 (a type of pollution particularly damaging to people's health) pollution particles per year. However, as so often happens, it is not as straightforward as it first appears, because scientists at York University discovered that if the temperature goes over 35°C London planes, and other trees such as willow and oak, start emitting large amounts of Volatile Organic Compounds (VOCs.) One of these, isoprene, combines with Nitrogen Oxide from car exhausts to produce dangerous ozone which is particularly a health risk for the elderly, children, and asthmatics. It also causes birds to have physiological changes including shortening their cilias (which filter air into their lungs) which leaves them more prone to disease. Summers in the United Kingdom are becoming hotter due to climate change, so it is a concern. However, the advantages of trees like London planes arguably outweigh the disadvantages because they are a hardy tree which can withstand high winds and disease, and they help to reduce flooding and regulate urban temperatures, and are able to tolerate poor, compacted soils.

Another tree which grows well in urban environments is the sycamore. It is not a native to the United Kingdom. No one knows how it got here. Some people think the Romans brought it over, whilst other people think it was brought back by knights returning from the Crusades, or by monks returning from pilgrimages around Europe. It has colonised Britain easily, especially in urban areas, because it is resistant to air pollution and salty air, two things which affect a lot of trees in urban environments, and its 'helicopter' seeds spread on the wind.

Some urban trees and shrubs are not so resistant to pollution and disease. Box wood is the heaviest native timber. It was used to make musical instruments. It is toxic but for centuries the steroidal alkaloids and flavonoids it contains have been used to treat a range of diseases including syphilis, leprosy and H.I.V. Lawson cypress was introduced to the United Kingdom in the mid-19th century and is named after the Scottish plant collector, Charles Lawson. Like box, it was used to make musical instruments.

When I was a solicitor my colleague showed me a double filing cabinet full of files and told me that they were the firm's bread and butter. They all contained files about border disputes, and in most cases it involved one tree- leylandii. I know from personal experience what a nuisance these trees can be. They were accidentally created in Wales in the mid-19th century when two North American cypresses cross bred. Leylandii grow remarkably high, up to 131 feet (40m), very quickly, blocking out light.

Another popular component of urban hedges is the privet. I have one growing around the boundary of my house. They have very pretty flowers, but their black berries are very poisonous but very popular with many birds (the blackbirds in my garden particularly like them.) The one in my garden is the garden privet, which was introduced from Japan, but wild privet also grows in the United Kingdom. One of the species which relies on privet is the privet hawk-moth, a striking moth with a pinkish body and long narrow wings.

In the town where I live there are many cherry trees which look particularly beautiful in the spring when their blossom is briefly in flower. Cherries were seen as the first fruit of Paradise in many Renaissance paintings. They produce melatonin which helps combat insomnia and their antioxidant and anti- inflammatory properties are currently being studied.

Another tree which is popular in urban streets and parks is the horse chestnut tree. It originates from the Balkans and became popular in this country in the 18th century. No-one knows how it got its name. Some people suggest that its nut, known as a conker, look like a horse's eye, whereas the Oxford English Dictionary suggests that conkers were used to treat respiratory disease in horses, although I have not found any evidence that this is true.

When I was a child, I was conker champion for two years running at primary school. The secret to my success was firstly living near a row of horse chestnut trees, so being able to select the best conkers before most of my classmates had had chance to have a look for any and dipping the conkers in vinegar and placing them in the airing cupboard for a week. Maybe I should have entered the World Conker Championships at Ashton in Northamptonshire, although I suspect they would not have allowed the vinegar. I had always thought that the game of conkers could be traced back centuries, but according to Fiona Stafford in her interesting book, *The Long, Long Life of Trees*, like many other things, it was the Victorians who first came up with the idea.

Another myth which Fiona Stafford dispels is that conkers deter spiders. For years I have been putting conkers in the corners of the room during Autumn in an effort to save myself from the huge spiders which inhabit my living room that time of year, but it seems that conkers cannot save me. In 2009 the Royal Society of Chemistry laid down a challenge for someone to prove that conkers deterred spiders but despite people devising numerous scientific experiments no-one was able to show that spiders were remotely deterred by a conker in the corner of the room.

Why are urban habitats important?

Urban habitats only cover 7% of the United Kingdom's land mass, but more than 80% of the British population live in towns and cities. The Industrial Revolution saw a huge cultural shift from a mainly rural population to a mainly urban one in the space of two or three generations. However, well off Victorians did not forget the countryside. Many natural history societies and environmental charities were created during his time and following the railways and statutory holidays, by the late 19th, early 20th century pursuits such as birdwatching, rambling, or cycle rides around the country were popular. The author and birdwatcher Stephen Moss says, "By the closing decades of the 20th century the countryside of our grandparents' childhood had, to all intents and purposes, ceased to exist. In its place was a green desert: clean, efficient, yet in most places totally devoid of wildlife." However, at the same time there has been an increase in what the nature writer Richard Mabey describes as the 'unofficial countryside', the urban brownfield sites, railway embankments, grass verges, canal towpaths, urban gardens, and parks. Two out of three of us feed the birds in our garden, and charities like the R.S.P.B and Wildlife Trusts have greater memberships than political parties. As Stephen Moss says, "...urban Britons are just as connected- arguably sometimes more connected – than their rural neighbours...far from losing our passion for nature, we city dwellers need it more than ever."

Remarkably, scientists think that half the plants which grow in the Northern Hemisphere are found in urban areas. Urban habitats are not just good for wildlife (providing an abundance of food, habitats, and much needed green corridors), they are also incredibly important in terms of humans' health and well-being. Studies have shown that being in a green environment has significant impact on physical and mental health. King's College, London developed an app called Urban Mind to examine how natural features affect mental wellbeing and discovered that being outdoors and seeing trees and the sky, and listening to birdsong, were associated with high levels of mental wellbeing.

Trees and other plants in urban areas provide shade and shelter, regulate urban temperatures, reduce flooding, and help reduce noise and air pollution.

Urban environments provide richly biologically diverse habitats. Greater London is the largest city in Western Europe with a population of over 8.6 million, but almost half of it is green or blue space, and it is the most species diverse area in the whole of the United Kingdom. More than 13,000 species call Greater London home including 1500 plants, 300 birds and the Thames, which about seventy years ago was so polluted nothing lived in it, now has 125 species of fish. The Greater London region has nationally important populations of species including stag beetles, greater-yellow rattle, and black redstarts.

Children who live near urban green spaces have been shown to have better concentration and motor skills. Access to green spaces also helps combat obesity and reduce anxiety and depression. It also encourages people to interact better with their neighbours, increasing a sense of community and combating loneliness, an especially big problem due to the covid-19 pandemic.

Threats to urban wildlife

Species are increasingly moving from rural to urban areas. Peregrine falcons, red kites, foxes, badgers, deer, and gulls are just some of the species thriving in urban areas. However, this is not the whole story. Out of the 529 urban species in the United Kingdom, 37 are at risk of extinction.

There are many reasons why urban wildlife is under threat, but a big cause is pollution – light, noise, and air.

In urban areas the night skies are sometimes thousands of times lighter than they were 200 years ago. This has contributed to a huge reduction in insect numbers, as well as confusing migratory birds, leading them to collide with tall buildings. Light pollution particularly affects the circadian rhythms of nocturnal animals which either causes them to avoid the light, thus affecting their hunting or reproductive patterns, or making them excessively attracted to the light. Light pollution is significantly impacting on the predator/ prey balance which has a huge effect on our wildlife populations.

Light pollution also affects the production of melatonin, a hormone produced by the pineal gland at night. A shortage of melatonin affects humans' sleep patterns which leads to an increase in anxiety and stress and causes many other health problems. There is a link between decreased melatonin and cancer. Blue light from mobile phones and computers and L.E.D. lights are particularly bad for melatonin levels so switch off at night and do not have your mobile phone in your bedroom at night. Animals also need melatonin and a reduction of it also affects their sleep patterns and bodily functions, affecting their night vision, breeding, and appetite.

Noise pollution means that birds and animals cannot communicate with one another, and this affects breeding and feeding. Christopher Templeton at the Pacific University gave zebra finches a series of tasks which mimicked looking for food in the wild. Then they carried out the tasks again whilst listening to a soundtrack of traffic in a semi-rural area. He discovered that the finches took twice as long to carry out the tasks whilst the soundtrack was being played. Another study by Cambridge University has revealed that traffic noise reduces the mating success of two-spotted crickets by 20%.

A 2016 study found that air pollution has increased by 8% globally in just five years. The worst affected areas of the world are in Asia and the Middle East, but air pollution is also a huge problem in the United Kingdom, resulting in more deaths per year than asthma, cancer and heart disease combined. Since 2010 nitrogen oxide has been well above legal levels, mainly due to diesel vehicles. Although all living things need nitrogen, one of the consequences of too much nitrogen in the environment is that nitrogen-loving plants, such as nettles, brambles, and cow parsley, take over environments, meaning that more vulnerable plants, mosses, and lichens cannot compete. Another consequence is that it makes trees grow too quickly which makes them more susceptible to disease.

Plantlife estimates that about 63% of sensitive habitats in the United Kingdom have too much nitrogen. This has a huge impact on ecosystems. Kevin Hicks at Stockholm Environment Institute believes that air pollution is one of the causes of the sixth mass extinction which is currently being experienced worldwide.

It is not just nitrogen oxide which is harmful to wildlife and humans alike. Carbon monoxide affects birds' breathing and central nervous systems and is fatal to humans if absorbed in sufficient quantities. Sulphur dioxide, a by-product of burning coal and oils, damages birds' immune systems, and leads them to produce fewer eggs. Ammonia is mainly from agricultural run-offs, but also from catalytic convertors. More than 80% of England's land mass has critically high levels of ammonia. It causes acidification, damaging plants and making them more susceptible to pests, disease, and adverse weather conditions. Air pollution affects birds and animals' habitats and food sources, which in turn affects their reproduction and makes them more susceptible to pests and disease. It also affects plants, making tree canopies thinner, thereby leaving birds more visible to predators, and reducing the tree's ability to regulate temperatures and reduce flooding.

Urban trees face a lot of problems including salt from the roads, soil compaction and, worse still, tarmac, utility companies trimming their roots, and having posters nailed to them, which allows in pests and disease.

The total area of gardens in the United Kingdom is equivalent to the size of Suffolk, and they are important habitats and green corridors. However, there has been a growing trend in recent years to pave over gardens and remove plants and shrubs. A third of London's 3.8 million gardens have been paved over. A third of gardens nationally do not have any plants in them. This has had a huge effect on many species but especially pollinators such as bees and butterflies, and insects such as worms and beetles. The reduction in insects have affected the populations of birds and mammals which feed on them. The R.S.P.B. notes that house sparrows, starlings, and song thrushes have reduced by 50% in the last 25 years partly because of the reduction in food and habitat loss.

The current government's green credentials are questionable. Amongst other things the relaxation of planning laws, taking the power away from local councils and residents, has been looked at in horror because it is worried that developers will be able to build on green spaces without local people being able to stop them. At the time of writing more than three thousand new homes are being proposed near the Knepp estate in Sussex, one of the most important rewilding sites in the United Kingdom, home to many rare species. If it goes ahead this will remove important green corridors which environmentalists now realise are imperative to the recovery of our natural environment because they allow wildlife to move between habitats and do not restrict them to small, fragmented areas.

What is being done

There has been a proposal to create the Greater London National Park. Countries like America and Canada have urban national parks, but if it goes ahead the Greater London National Park will become the first national park in the world to encompass an entire city. It is hoped that it would provide some legal protection for London's multitude of green spaces but would also encourage its human inhabitants to do more to help nature in their gardens, streets, and parks. As part of the plan, it is hoped that reservoirs could become wetlands and that industrial sites could become sites of special scientific interest.

Following the covid -19 pandemic many individuals and organisations are calling for a green revolution. During lockdown wildlife benefited from less traffic and industry and many people had time to watch the wildlife in their parks and gardens and relished their daily walks in green spaces. Charities such as the National Trust argue that this is an ideal time to invest in urban green infrastructure. The charity has called for a £5.5 billion green infrastructure fund to allow more people access to green spaces. Currently 440,000 people live in the most deprived urban areas which are 'grey deserts' devoid of trees and green spaces. Twenty per cent of the poorest households do not have a car and public transport is so patchy in many areas that it is difficult for urban dwellers without a park or garden to access the countryside. This has a huge effect on human health. The National Trust argue that the fund would lead to £200 billion savings because of better physical and mental health. They estimate that a third of the British population would benefit from the fund.

The National Trust propose that the fund could be spent on connecting green spaces, creating street parks, and upgrading parks, as well as planting urban forests. All over Britain urban councils are trying to improve green spaces. In Plymouth they are going to create green boulevards and public squares. In the West Midlands they plan to create a regional park which would encompass several towns and cities. Exeter's Wild City scheme by the Council and various partners includes the first swift tower, awards for the best wildlife garden, and a nature education programme.

In Todmorden on the Lancashire/Yorkshire border two ladies had a fantastic idea about fourteen years ago. Pam Warhurst and Mary Clear began to plant fruit, vegetables, and herbs around the town. People can help themselves to it. There are now more than seventy beds around the town. Soft fruits are grown near the local care home so the residents can help themselves to easy- to- eat fruit, herbs are grown outside the local butchers to compliment the meat sold, medicinal plants are grown near the health centre. Each business or organisation adopts a bed and with a team of volunteers cares for it. Three-quarters of the residents of the town are now growing their own food. It has had a hugely positive effect on the town, reducing anti-social behaviour and encouraging new businesses to open in the town. The Incredible Edible movement has spread to 120 places in the United Kingdom, including York. A under -used side street in Islington has been turned into a street park with an edible walkway.

I went to university in Nottingham. It has many beautiful buildings, but the Broadmarsh shopping centre is not one of them. It is a 1960s monstrosity. I think that most people in Nottingham were happy to hear that it was to be demolished. Initially houses were to be built on the site, but the housing firm has gone bankrupt because of the covid-19 pandemic. Rather than cause a problem this has given Nottingham Council a wonderful opportunity. Nottingham Wildlife Trust has proposed that the area, in the centre of Nottingham, should be transformed into a public green space consisting of six acres of wetlands, meadows, and woodlands and green corridors to nearby Sherwood Forest. The proposal has a lot of public support, and I am one of the many people who hope that it comes to pass.

Whilst many cities are taking steps to improve their green spaces, Bristol is Britain's greenest city. Since 2015 the Council, in partnership with other organisations, such as the Avon Wildlife Trust, have created community orchards and grasslands and meadows, put up bird boxes and shelter for wildlife, planted nectar-rich plants, and provided water. It also has beehives dotted around the city. Many Bristolians have taken part, whether it is improving their gardens or educating school children in the importance of looking after wildlife. Bristol now has 87 wildlife sites and has created some important green corridors. It has also created five urban farms, as well as communal allotments on former brownfield sites.

The National Capital Committee has suggested that planting 965 square miles of woodland close to urban centres would provide an economic benefit of £550 million in recreation and carbon sequestration. There are already some 'tiny forests' in the United Kingdom. This is an idea which started in Japan about fifty years ago and involves lots of trees in a small space. In Bristol they are planting a 'tiny forest' the size of a tennis court which will have 600 native trees including the exceedingly rare black poplar. Urban trees absorb carbon more quickly than rural trees, but they also grow faster and die younger than their country counterparts.

Trees for Cities is an organisation which have planted more than a million urban trees in almost thirty years including a wood in North-East London and a horticultural therapy centre at the Maudsley Hospital in London.

Urban planners are beginning to incorporate biophilic designs into urban environments. These are designed to help health and wellbeing such as having a lot of plants and trees near a hospital to help promote faster healing. The idea of biophilic cities was first launched in 2013, and Birmingham became the first biophilic city in the United Kingdom. The idea of a biophilic city is to have an abundance of nature in close proximity to residents; value, protect and restore biodiversity; make it easier for people to access green spaces, provide learning opportunities in a natural environment, and give residents a closer connection with nature through parks, natural history museums, and other wildlife sites. Natural materials should be incorporated into buildings. Living walls and green roofs look nice and are good for humans and nature. A relatively new concept is vertical farming which is the cultivation of vegetables and edible plants on a vertical surface. One acre of vertical farming is equivalent to ten-twenty acres of conventional farming. As well as providing food and regulating the temperature of buildings, and improving air quality, these green walls also provide valuable habitats.

Another way of growing vegetables is increasing in popularity. Aquaponics combines growing vegetables and keeping fish. The fish poo adds nutrients to the water, which helps the vegetables to grow, and the plants filter the water, improving the water quality for the fish.

During lockdown in 2020 a study of urban, white-crowned sparrows in San Francisco were found to sing more quietly because of the reduced noise pollution. At the time of writing, we are still in the grip of the pandemic, but it is thought that in the post pandemic world some people will chose a new form of working. Instead of going to the office, some firms have realised the benefits of staff working remotely from home. In addition, it is believed there will be a reduction in air travel due to the pandemic and climate change. If these changes do happen it will have a huge effect on noise pollution. Academics are also looking at changes to cars and aircraft to make them quieter. In the United Kingdom the government wants to phase out diesel and petrol cars in the next decade. This would also have a positive effect on noise pollution because electric cars are quieter, as well as not emitting dangerous chemicals into the air. However, at the time of writing, there does not seem to be any plan on how to develop the infrastructure required for widespread electric cars.

What you can do to help urban wildlife

If you have a cat think about putting a collar on it or keeping it inside during the bird breeding season. Cats are one of the main predators of songbirds, and also predate bats and other mammals.

Think about what to plant in your garden. The Royal Horticultural Society has discovered that the hairy-leafed cotoneaster franchetii soaks up 20% more pollution compared to other shrubs if planted near busy roads. Privet, bay laurel, box, yew, holly, silver birch, elders, conifers, wallflowers, and ivy have also been shown to be effective at soaking up air pollution. If you live on a busy street, consider having a hedge made out a mixture of some of these. It will also help absorb noise pollution. However, it is important to have the right species in the right places. Large trees on busy streets are good at regulating temperatures and reducing flooding but acts like a roof, trapping air pollution at ground level, so if you are in a busy urban area plants trees which do not grow as high. Hawthorn and privet hedges are also effective at soaking up rainwater and reducing the risk of flooding and ivy clad buildings help insulate the building in winter and cool it in summer.

Trees are particularly effective when removing parciculate matter (P.M.s) which are as small as a fifth of the width of the human hair, and which globally kill eight million people per year. Trees help disperse the P.M.s or catch them in their leaves. Rough, rugged, hairy plants are best at trapping the P.M.s so think about planting some in your garden.

An easy way to reduce light pollution is to turn off lights when they are not needed and to use shields on outdoors lights. The University of Exeter found that the number of species affected by L.E.D. lighting was greatly reduced when lights were dimmed by 50% and turned off between midnight and four in the morning.

One of the most important things you can do is to make your garden more wildlife friendly. Make your garden more hedgehog friendly, plant lots of nectar rich plants for pollinators, and leave scruffy areas for amphibians, reptiles, and small mammals. Think about planting different coloured plants which flower at different times so that you are helping different pollinators, and in doing so helping bats who feed on moths. There are resources listed at the end of this book. Why not build a bug hotel or a hedgehog house? It is fun to do, and it helps nature. Also make sure that you have gaps in hedges and fences so that hedgehogs and other animals can use the garden as a wildlife corridor. It only needs to be a small gap, but it will make a big difference. Put up feeders and bird boxes for the birds and let all of, or part of, your lawn grow long and wild. Plantlife have a No Mow May campaign. An unmowed lawn supports 4000 bees as well as many other pollinators and insects. I leave part of my lawn long. I was delighted to find a colony of grasshoppers living in it this year. If you do not want to leave it completely overgrown, reduce the number of times you cut it to once a month, and try adjusting the cutters so that it does not cut the smaller plants such as daisies, clover, and dandelions which provide a lot of nectar for pollinators and look pretty. Also do not weed regularly. A weed is just a plant after all. Plants such as clover add nitrogen to the lawn, if left to grow, making the lawn healthier. Do not use fertilizer or peat. Instead get a compost heap or see if your local riding school has some manure going spare. There's also peat free soil sold at most garden centres.

The Wonders of the Wild Places

One of the best things you can do is to create a pond but remember to keep the edges rough, with lots of stones and cover for amphibians. There is advice at the back of this book about creating a garden pond.

The Friends of the Earth have a Ten Times Greener campaign. It involves neighbours working together to make their streets greener and to grow plants for food and wildlife. The Friends of the Earth provide 'postcode gardeners' to help and advice. More information is at the end of this book. Alternatively, you could set up an Incredible Edible movement in your community. I am going to try and do this in my town. I will ask businesses to adopt a planter and see if we can get some of the younger members of the community to volunteer to care for them.

Naturalists such as Chris Packham and Sir David Attenborough are currently urging Britons to plant one billion seeds in order to combat climate change and the B.B.C's Countryfile programme has launched a Plant Britain campaign. Details are in the Appendix at the back of this book. I am in a community group which helps plant trees, bulbs, and plants around the town where I live. We have recently created a community friendship garden, and also have a community orchard. We help to maintain the woods and parks in the town.

It makes the town a nice place to live for humans and wildlife alike. If you have something similar in your area, think about joining it. If not, why not set one up?

Hope for the future?

As I have shown in this book nature is in trouble. Frighteningly for me, it is not just one habitat, genus or species which is affected- from woodland and wetlands to heathland, from the highest mountain to the deepest ocean, urban areas to farmland, nature is in crisis. Equally worrying is the multitude of threats that nature faces- including habitat loss, invasive species and pests, and climate change. Worse still, it is not just this country which is affected. Worldwide, habitats and species are being wiped off the planet. One million of the eight million species of animals and plants face extinction. Across the world there has been a 60% decline in just fifty years. There have been mass extinctions before, but what concerns scientists is the speed at which it is happening. A 2019 United Nations report by five-hundred leading scientists concluded that we are losing biodiversity at a rate not seen before in human history. This not only threatens wildlife, but also threatens humans. We need healthy soils, seas, and biodiverse habitats for our health and ongoing existence too. Nature, and humanity, has never been in such a precarious state.

However, this is not the end of the story. It does not have to be this way. We can change the narrative, but because of inaction by governments all around the world for the last few decades we now do not have much time to change things around. There will be a tipping point – a point of no return- if we do not make big changes to the way we live in the next decade.

For the sake of all the species that we share this planet with, as well as for ourselves and our children and grandchildren we must make those changes. As one of the leading climate experts, Professor Robert Watson, says, "If we do not act now the youth of today and tomorrow will look at this generation and say' what were you thinking?'"

All over the world individuals and organisations are working hard to help nature and restore habitats and biodiversity. I hope you have got inspiration from reading this book and that you will get involved- whether it is making your garden more wildlife-friendly, cleaning your local beach, or volunteering with your local environmental group or with a wildlife charity- or better still doing more than one thing. However, much more is needed. We can only save nature if things are done on a bigger scale and more quickly. For this to happen governments across the world must start taking the threats from nature loss and climate change seriously, and invest in making cities greener, seas cleaner, protecting and planting forests, and reducing chemicals used in agriculture. We must break with our dependency with fossil fuels and plastic. We must start seeing the value in the natural world- not just its beauty but also how it affects human health and wellbeing. We can no longer pollute and destroy; we must plant and nurture. If given it the chance nature can bounce back, but there has to be the intention to make a difference and across the world I am not seeing this at the moment.

Our own government claim to have green credentials but at the same time as championing tree planting programmes they allow major infrastructure projects, such as HS2, to decimate ancient woodland, a valuable and irreplaceable habitat. Whilst claiming to be on the side of nature in recent months the government has not spoken out about a proposed new coal mine in Cumbria, has allowed new nuclear power stations to be built, has relaxed planning laws to make it easier to build on green land, has failed to prosecute companies who have polluted rivers, and allowed a pesticide, known to severely threaten bees and pollinators to be used. Only 7% of our woodland is in a good condition, only 3% of our meadows survive, our seas, rivers and air are polluted, more than 40% of our Sites of Special Scientific Interest are in poor condition, over intensified agriculture has damaged our farmland. It is difficult sometimes not to be gloomy.

However, researching this book has given me some hope. People are making a difference. We know what the problem is, and we have everything we need to fix it, but we must work together as a species to protect our planet for ourselves and for every other species which calls it home. We must start putting pressure on our governments to put their money where their mouths are. We must not allow them to ignore the issue any longer. Big changes must be made, and they must be made now.

Some changes require seismic shifts in thinking – such as moving away from fossil fuels – although we already have the alternatives –

solar, wind and water power. However other changes just require us to do nothing. There is a growing rewilding movement in Britain and across the world. The idea is to allow areas to revert to the wild. Stop ploughing and spraying, stop trying to fight the wildness, let the land go back to the way it was. Rewilding Britain aims to rewild 1158 square miles of Britain and create three marine areas by 2030. Numerous rewilding projects are ongoing including Wild Ken Hill Farm in Norfolk. The Trees for Life organisation is rewilding areas across the United Kingdom. The Knepp estate in Sussex has already shown the difference rewilding can make in a relatively short period of time.

Rewilding is just one solution. In this book I have shown that lots of different solutions are required- whether it is planting more seagrass and seaweed in the sea, or banning chemical pesticides, or restoring habitats, or using technological advances to reduce pollution. However, all these projects require funding and that is why governments must invest in a green revolution to save our wildlife and our planet.

The recent covid pandemic has shown what things could be like. During lockdowns, pollution was reduced, nature took over urban areas, we spent more time tendering our gardens and many people became aware of the beauty and health benefits of nature. The covid pandemic is a perfect opportunity for humans across the world to pause, reflect, and reset. As countries come out of the pandemic it is the perfect time for governments to invest in a greener future. It is not just for aesthetic or even moral reasons, where nature flourishes people flourish too. More green spaces and biodiversity means less health problems and crime. It means safer and cleaner cities and more sustainable farming. We must not waste this opportunity.

Can you imagine a world without a bluebell? Or birdsong? Or whales? Or trees? No, neither can I, but that might become a reality for future generations if we do not act now. So, I urge you to do what you can – plant, campaign, change your energy, reduce your plastic consumption, remember the natural environment when you vote- please do everything you can for nature. I am a firm believer that small changes make a big difference. One person's actions can affect others. Join your neighbours to make your gardens more wildlife friendly, encourage your colleagues and families to live more sustainably, ensure our politicians are not only listening but also acting to improve the natural environment and take steps to limit climate change. Together we can make a difference, but we must act now.

What better legacy can we leave on this earth, than knowing that because of our actions, for centuries to come people can enjoy the wonders of the wild places.

Appendix

This gives you apps, websites, and books where you can obtain further information. It is not exclusive, but this will give you a starting point if you want to learn more.

IDENTIFICATION

- **Flowers** – Apps- Picture This
  - FlowerChecker
  - PlantNet
  - PlantSnap
  - Wildflower Id
  - Flowers of Britain
  - Google lens

  - Websites- www.plantlife.org.uk
  - Books - Wild Flowers (Collins Gem) – Highly recommended and small enough to put in a rucksack or pocket.
  - Herbs for Cooking and Health (Collins Gem Guide)
  - Wild Flowers (Collins Nature Guide)- Highly recommended. Easy to use.
  - Herbs and Healing Plants (Collins Nature Guide)

- **Trees -** Apps- The Woodland Trust has a free tree identification app. See www.woodlandtrust.org.uk
  - www.treeguide.co.uk
  - Google lens
  - Books- Trees (Collins Gem)

- **Butterflies and Moths-** Apps- iRecord Butterflies- see **https://butterfly-conservation.org**
  - Books- Butterflies and Moths- (Collins Gem)

The Wonders of the Wild Places

- Website- **www.butterfly-conservation.org**
- www.uk.moths.org.uk
- www.woodlandtrust.org.uk
- www.butterflyidentification.com

- **Birds**
- Apps-Birds of Britain Pro
- Collins British Bird Guide
- Chirp!
- British ID- British Isles identification guide
- eGuide to British Birds
- iBird UK Pro
- Warblr
- Birdtrack
- **Books**- RSPB Handbook of British Birds
  RSPB Pocket Guide to British Birds- Easy to use
  Birds (Collins Gem)

- **Insects**
- Collins Gem
- www.buglife.
- www.uk.beetles.co.uk
- www.bumblebeeconservation.org
- www.bna-naturalists.org

- **Reptiles**
- www.arc-trust.org
- www.wildlifetrusts.org.uk

- **Amphibians**
- www.arc-trust.org
- www.wildlifetrusts.org.uk
- www.froglife.org

- **Mammals**
  - www.mammal.org.uk
  - www.bats.org.uk
  - www.wildlifetrusts.org

- **Fungi**
  - www.woodlandtrust.org
  - **www.nhbs.com/blog/the-nhbs-guide-to-fungi-identification**

- **General Apps**
  - iNaturalist
  - Google Lens
  - Nature Finder (Wildlife Trusts)
  - iRecord

The I- Spy series is great for children and makes looking for nature fun. The Usborne books are also highly recommended for children.

ORGANISATIONS

www.nationaltrust.org.uk – Information about volunteering opportunities and membership

www.wildlifetrusts.org- Great for identification, information about campaigns, membership, and volunteering.

www.rspb.org.uk - Great for identification, information about campaigns, membership, and volunteering.

www.woodlandtrust.org.uk Great for identification, information about campaigns, membership, and volunteering.

www.plantlife.org.uk - Great for identification, information about campaigns, membership, and volunteering.

www.barnowltrust.org.uk – Adopt a barn owl, Volunteering opportunities.

www.bto.org (British Trust of Ornithology) - Great for identification, information about campaigns, and volunteering.

The Wonders of the Wild Places

www.swift-conservation.org – Information

www.buglife.org.uk - Great for identification, information about campaigns, and volunteering.

www.wwf.org.uk- Adoption, information about campaigns, membership, and volunteering.

www.butterfly-conservation.org- Great for identification, information about campaigns, membership, and volunteering.

www.wwt.org.uk (Wildfowl and Wetlands Trust) – information about campaigns and information about ponds and gardening.

www.arc-trust.org- Great for identification, information about campaigns, membership, and volunteering, adoption, advice.

www.hawkandowltrust.org- Great for identification, information about campaigns, membership, and volunteering.

www.therivertrust.org -Information about campaigns, membership, and volunteering.

www.ypte.org.uk – (Young Person's Trust for the Environment) – Information for teachers and parents.

www.bumblebeeeconservation.org - Great for identification, information about campaigns, membership, and volunteering.

www.friendsoftheearth.uk - Information about campaigns, membership, and volunteering.

www.bats.org.uk (Bat Conservation Trust) Adoption, identification, information about campaigns, membership, and volunteering.

www.treesforcities.org information about volunteering and current projects

www.hedgelink.org.uk information and resources

www.hedgelaying.org.uk information and resources

www.eagleintroductionwales.com volunteer opportunities

www.northyorkmoors.org.uk information about projects and volunteering opportunities

www.northumberlandnationalpark.org.uk information about projects and volunteering opportunities

The Wonders of the Wild Places

www.cairngormsconnect.org .uk - information about projects and volunteering, resources for schools

www.cairngorms.co.uk Information about projects and volunteering

www.snowdonia.gov.wales Information about projects

www.scottishwildlifetrust.org Information about projects, membership, and volunteering

www.froglife.org Information

www.wildottertrust.org

www.scottishbeavers.org.uk

www.waterways.org.uk

www.naturalresources.wales

CAMPAIGNS

Billion Seed Challenge

Plant Britain – see www.plantbritain.co.uk

No Mow May- Plantlife

Ten Times Greener – Friends of the Earth.

Check, clean and dry – to prevent the spread of invasive species. – www.therivertrust.org and www.canalrivertrust.org.uk

Thirteen ways to Save water www.friendsoftheearth.uk/13-best- ways -to- save- water

Plantlife's Every Flower Counts- www.plantlife.org.uk/everyflowercounts

Countryfile's Plant Britain www.plantbritain.co.uk

Plantlife – No Mow May

Plantlife- Making Meadows

PROJECTS

MOREHedges – www.woodlandtrust.org.uk
Magnificent Meadows – www.magnficientmeadows.org.uk
Snakes in the Heather -www.arc-trust.org
Great Heath Project – www.dorsetwildlifetrust.org.uk

Fix the Fells- www.fixthefells.co.uk
PEATlife -www.northpennines.org.uk
Moors for the Future www.moorsforthefuture.org.uk
Dynamic Dunescapes- www.dynamicdunescapes.co.uk

HOW TO GUIDES
Creating and maintaining hedges
CPRE, A Little rough around the hedges -Lots of information about hedges and species

Information about how to create a pond
www.arc-trust.org
www.wwt.org.uk/ gardening for wetlands/how to build a wildlife pond.
www.froglife.org – Information about pond creation and wildlife gardening.
www.naturehood.uk/wildlife-garden/ build-a-pond
www.freshwaterhabitats.org.uk
www.rspb.org.uk
www.wildlifetrusts.org

Information about how to make a wormery
www.schoolgardening.rhs.org.uk

Information about making a wildlife meadow
www.friendsoftheearth.uk

How to attract hedgehogs into your garden
www.hedgehogstreet.org

The Wonders of the Wild Places

References

INTRODUCTION
1. Moss, S. (2018) Mrs Moreau's Warbler. Faber
2. Mabey, R. (2016) The Cabaret of Plants. Profile Books
3. Cairoli, S. (9.15.18) The Disadvantages of Deforestation. Available at www. sciencing.com
4. BBC1, Extinction, programme aired 13.9.20
5. Shukman, D. (17.9.21) Climate change: UN warning over nations' climate plans. BBC. Available at www.bbc.com
6. State of Nature Report 2019
7. Tree, I. (2018)Wilding. Picador
8. www.jncc.gov.uk
9. Spencer, B. Barking up the wrong tree! 98% of us can't name five common species from looking at pictures of their foliage. Daily Mail. (2.5.14) Available at www. dailymail.co.uk
10. Stuart-Smith, S. (2020) The Well-Gardened Mind: Rediscovering Nature in the Modern World. William Collins
11. North York Moors National Park. www.northyorkmoors.org.uk
12. Harrabin, R. (28.9.20) Boris Johnson promises to protect 30% of UK's land by 2030. BBC. Available at www.bbc.com
13. Woodland Trust. (11.6.20) Disappointing planting figures in England still far below government targets. Available at www.woodlandtrust.org.uk/press-centre/2020/06/government-planting-figures
14. Agriculture Act 2020
15. Morelle, R. (13.6.22) Huge plan to map the DNA of all life in the British Isles. BBC. Available at www.bbc.co.uk/news/science-environment-61747729

IF YOU GO DOWN TO THE WOODS TODAY
1. Brown, Paul. England's woodlands growing to 1000-year record. The Guardian 22.11.01
2. Royal Forestry Society. www.rfs.org.uk
3. Wikipedia
4. Wright, J. (2016) A Natural History of the Hedgerow and ditches, dykes and stone walls. Profile Books.
5. Woodland Trust www.woodlandtrust.org.uk
6. Countryfile BBC1 Broadcast 12.7.20
7. National Trust. Available at www.nationaltrust.org.uk
8. www.newforestpa.gov.uk
9. Barkham, P. HS2 will destroy or damage hundreds of UK wildlife sites, says report. The Guardian (15.1.20) Available at www.theguardian.com

10. Tudge, C. (2005) The Secret Life of Trees. Penguin
11. www.treeguideuk.co.uk
12. Tree, I. (2018) Wilding Picador
13. www.visitsherwood.co.uk
14. www.royalparks.org.uk
15. Derwent, L and Gillespie, T.H. (No date) Nature's Wonders. Collins.
16. www.forestresearch.gov.uk
17. Wohlleben, P. (2016) The Hidden Life of Trees.. William Collins
18. Sheldrake, M. (2020) Entangled Life. The Bodley Head
19. www.countrysideinfo.co.uk
20. www.gocycle.com
21. Woollacot, E. (No date) The fungus and bacteria tackling plastic waste. BBC. Available at www.bbc.com
22. Plantlife. Available at www.plantlife.org.uk
23. Stafford, F. (2018) The Brief Life of Flowers. John Murray.
24. Countryfile. BBC1. Broadcast 24.4.21
25. Culpepper, N. (No date)Culpepper's Complete Herbal
26. Simons, P. Magical rainforests of Britain need of protection. The Guardian (27.1.16) Available at www.theguardian.com
27. www.statistica.com
28. The Hidden Life of Motorways. BBC2. Broadcast 12.11.21
29. Moss, S. (2018) Mrs Moreau's Warbler. Faber
30. Countryfile magazine. (No date) Pine Martens. Available at www.countryfile.com
31. Project Pine Marten. www.gloucestershirewildlifetrust.co.uk/project-pine-marten
32. Q.I. BBC2
33. Wildlife Trusts www.wildlifetrusts.org
34. Natural England. Review of Red Squirrel Conservation Activity in Northern England. 16.9.09
35. Rowlett, J. (26.1.21) U.K. government backs birth control for grey squirrels. BBC. Available at www.bbc.co.uk/news/science-environment...
36. Red Squirrel Survival Trust. www.rsst.org.uk
37. Iles, J. (15.8.20) New Research shows pine martens preying on grey squirrels, more than native red squirrels. Bromsgrove Standard. Available at www.bromsgrovestandard.co.uk
38. Hammill, J. Squirrels can be left-or right-handed and their orientation affects their intelligence. The Metro. (20.1.20) available at www.metro.co.uk
39. Couzens, D. (14.10.19) Red squirrels guide: how to identify and best places to see in Britain. Available at www.countryfile.com
40. Carrell, S. Scottish wildcats on the verge of extinction. The Guardian. (27.2.19) Available at www.theguardian.com
41. www.scottishwildcataction.org

42. Gibbs, J. (3.1.21) Scottish wildcats: 6 things you should know about the 'Highland tiger' and how it is fighting for survival in the 21st century. Countryfile Magazine. Available at www.countryfile.co.uk.
43. BBC. (22.4.21) Wildcats could return to England after 200 years. Available at www.bbc.co.uk
44. Springwatch. BBC2. Broadcast 10.6.20
45. www.scottishbadgers.org.uk
46. Springwatch. BBC2. Broadcast 25.7.20
47. RSPB. www.rspb.org.uk
48. www.field-studies.council.org/cuckoos
49. Barkham, P. Half the trees in two new woodlands planted by jays, study finds. The Guardian. (16.6.21) Available at www.theguardian.com
50. Jones, P.A. (9.5.15) Magpie. Available at www.haggardhawks.com/post/magpie
51. Willsy28. (6.9.17) Birds we named after ourselves. Available at www.birdsandtrees.net/2017/09/06/birds-we-named-after-ourselves
52. Jackdaw. (No date) Available at /www.word-detective.com/2008/07/jackdaw
53. Superstitious of Magpies: One for sorrow or one for joy? (13.3.22) Available at https://superstitionsaturday.com/2022/03/12/superstitious-of-magpies
54. One for sorrow: Magpie nursery rhyme. (11.11.20) Available at www.birdspot.co.uk/culture/one-for-sorrow-magpie-nursery-rhyme
55. Radio 4 in Four. (No date) 7 birds and their mysterious folklore. Available at www.bbc.co.uk
56. Autumnwatch. BBC2. Broadcast in 5.11.20
57. Smyth, R. (2017) A Sweet, wild note: What we hear when birds sing. Elliott and Thompson Ltd.
58. MacDonald Lockhart, J. (2016) Raptor: A journey through birds. 4th Estate.
59. Lewis-Stempel, J. (2014) Meadowland: The Private life of an English field. Black Swan
60. Barnes, S. (2016) The Meaning of Birds. Head of Zeus Ltd.
61. DEFRA. (2018) Tree Health Resilience Strategy.
62. Cheffings, C.M and Lawrence, R. (February 2014) A summary of the impacts of ash die back on UK Biodiversity, including the long-term monitoring and further research on management scenarios. Available at www.data.jncc.gov.uk
63. www.heartofenglandforest.org
64. Ivens, S. (2018) Forest Therapy: Seasonal ways to embrace nature for a happier you. Pratkus
65. Richardson, R. (2017) Britain's Wild Flowers: A Treasury of Traditions, Superstitions, Remedies and Literature. Pavilion Books

The Wonders of the Wild Places

HEDGES AND EDGES
1. DEFRA. (2018) Tree Health Resilience Strategy.
2. Buglife www.buglife.org.uk
3. Woodland Trust www.woodlandtrusts.org
4. Ancient and species rich hedgerows. Summary Action Plan. Doncaster Local Biodiversity Action Plan. (January 2007) https://dmbcwebstolive01.blob.core.windows.net/Default/Planning/Documents...
5. www.cornishhedges.co.uk
6. Hedgelink. Available at www.hedgelink.org.uk
7. www.stastica.com
8. Wright, J. (2016) A Natural History of the Hedgerow and ditches, dykes and stone walls. Profile Books.
9. Hooper, M, Moore, N. W., Pollard, E. (1974) Hedges. Collins.
10. Historic England. Available at www.historicengland.co.uk
11. Wildlife Trusts. www.wildlifetrusts.org.uk
12. www.treesforlife.org
13. Stafford, F. (2016) The Long, long life of Trees. Yale
14. A Little Rough Around the Hedges, CPRE
15. Mabey, R. (2016) The Cabaret of Plants. Profile Books
16. Rose Hip Information. (No date) www.gardeningknowhow.com/ornamental/flowers/roses...
17. Wikipedia
18. Interesting facts about holly. (No date) www.funfacts.com/interesting-facts-about-holly
19. Wiley. (13.12.12) Prickly holly reveals ability to adapt to genetics to environmental change. Science. Available at www.sciencedaily.com/releases/2012/12/1...
20. National Trust. Available at www.nationaltrust.org.uk
21. Culpepper, N (No date) Culpepper's Complete Herbal
22. Grigson, Geoffrey. (1962) The Shell Country Book. Dent.
23. Plantlife. Available at www.plantlife.org.uk
24. Simons, P. Plantwatch: the ingenious fly trap hiding in Britain's hedgerows. The Guardian. (19.5.21) Available at www.theguardian.com
25. Dry Stone Walls. Available at www.farmwildlife.info/how-to-do-it-5/field-boundaries/dry-stone-walls
26. Springwatch BBC2 Broadcast 25.7.20
27. Lewis-Stempel, J. (2014) Meadowland: The Private life of an English field. Black Swan
28. Derwent, L and Gillespie, T.H. (No date) Nature's Wonders. Collins.
29. Barkham, P. UK Roadsides on the verge of becoming wildlife corridors. The Guardian (27.9.19) Available at www.theguardian.com

30. Red clover: uses, benefits, side effects and more. (7.10.21) Available at www.hollandandbarratt.com/.../menopause/what-is-red-clover
31. History of the 4-Leaf clover and clover crafts. (17.3.20) Available at http://sctlandtrust.org/2020/0/17/history-of-the-4-leaf-clover...
32. UK's rarest plants at risk of extinction as road verges become their final refuge. The Telegraph (24.4.17) Available at www.telegraph.co.uk/news/2017/04/23/uk...
33. New Highway Agency Guidelines (September 2019)
34. Wood, C. (23.5.21) Coronavirus: Roadside wildflowers bloom under lockdown. BBC Wales. Available at www.bbc.com/wales
35. Greenfield, P. On the verge: a quiet roadside revolution is boosting wildflowers. The Guardian. (14.3.20) Available at www.theguardian.com
36. Jefferson, R. (No date) Understanding your verges. Natural England
37. Countryfile magazine. (No date) Guide to kestrels: how to identify and where to find them. Available at www.countryfile.com
38. Springwatch. Broadcast BBC2 on 5.6.20
39. Stoat- Top facts. Diet and habitat information. https://animalcorner.org/animals/stoat
40. Harrabin, R. (30.11.20) Brexit: Ministers unveil next steps in England's farming policy. BBC. Available at www.bbc.com
41. Storer, R. End mowing of verges to create huge wildlife habitat, says UK study. The Guardian. (26.5.21) Available at www.theguardian.com
42. Richardson, R. (2017) Britain's Wild Flowers: A Treasury of Traditions, Superstitions, Remedies and Literature. Pavilion Books

ON THE FARM
1. Sheffield University. (8.11.17) New land atlas reveals just 6% of the UK is built on. Available at www.sheffield.ac.uk
2. Lewis-Stempel, J. (2014) Meadowland: The Private life of an English field. Black Swan
3. Plantlife. Save our magnificent meadows.
4. Coles, J. (3.7.15) Why wildflower meadows are so special. Available at www.bbc.com.
5. Plantlife www.plantlife.org.uk
6. Wildlife Trusts www.wildlifetrusts.org
7. Lewis-Stempel, J. (2019) The private life of the hare. Doubleday
8. Barkham, P. Hopping for the best. Innovative new warrens set to stem decline of rabbits. The Guardian (27.11.21) Available at www.theguardian.com
9. Wikipedia
10. www.rabbitmatters.com
11. Collins, F. (20.1.22) Guide to rabbits and hares: what's the difference, where to see and species history. Available at www.countryfile .com
12. Bott, Adrian. The modern myth of the Easter bunny. The Guardian (23.4.11) Available at www.theguardian.com

13. Winick, S. Ostara and the Hare: not ancient but modern as some sceptics think. Folklore Today. Library of Congress. (28.4.16) Available at www.blogs.loc.gov
14. Barkham, P. Hare deaths raise fears rabbit virus has jumped species. The Guardian (27.8.21)   Available at www.theguardian.com
15. Bradford, A. (2015) Facts about moles. Available at www.livescience.com
16. Grigson, G. (1962) The Shell Country Book Dent.
17. Plantlife. The Good verge guide (2016)
18. McNab, C. The history of the Remembrance poppy. The Independent.   (10.11.14) Available at www.independent.co.uk
19. BBC Radio 4. (8.11.20) Natural Histories: Poppy. Available at www.bbc.com
20. Tree, I. (2018) Wilding Picador.
21. India Water Portal (No date) Earthworms for organic farmers. Available at www.indiawaterportal.org/sites/default/files/iwp2/earthworm...
22. Springwatch. BBC2. Broadcast 28.5.20
23. RHS Campaign for School Gardening. (No date) Worm fact sheet. Available at https://schoolgardening.rhs.org.uk/Resources/Info-sheet/worm-fact-sheet
24. Pidcock, R. Global warming: how worms are accelerating climate change. The Guardian (5.2.13) Available at www.theguardian.com
25. Yirka, B. (16.10.13) New study suggests earthworms sequester more CO_2 than they release. www.phys.org/news/2013-10-earthworms-sequester-co2.html
26. RSPB Farmland Bird Indicator
27. Barnes, S. (2016) The Meaning of Birds. Head of Zeus Ltd.
28. RSPB www.rspb.org.uk
29. Wimpenny, Dr. J. (No date) Crows have a reputation for being smart, but are they actually clever? Discover Wildlife magazine. Available at www.discoverwildlife.com
30. Corvidae- Stone the crows! (2.4.17) Available at www.pitchcare.com/news-media/corvidae-stone-the-crows.html
31. The Old Farmer's Almanac, 29.1.19, How birds predict the weather- weather proverbs about birds, Available at www.farmersalmanac.com
32. Spencer, H, As the crow flies; corvids in lore and legend, Folklore Thursday. (27.8.20) Available at www.folklorethursday.com
33. Rhodes, C. (2014) An Unkindness of ravens: A book of collective nouns. Michael O'Mara Books.
34. Nicholls, H, (8.4.15) The truth about magpies. BBC Earth. Available at www.bbcearth.com
35. Amazing facts about pigeons. (No date) Available at https://onekindplanet.org/animal/pigeon...
36. 21 Amazing facts about pigeons. (No date) Available at www.pigeoncontrolresourcecentre.org/html/amazing-pigeon-facts.html
37. Smithsonian Institute (4.1.21) STEM Visions. How do birds navigate? Available at www.ssec.si.edu.

38. Lund University. (6.4.18) How birds can detect the earth's magnetic field. Science Daily. Available at www.sciencedaily.com/releases/2018/04/180406091756.htm
39. Lee, J. J., (30. 1.13) New theory of how homing pigeons find home, National Geographic, Available at www.nationageographic.com
40. Bird, D; Freeman, R; Guilford, T ;Meade, J; et al, Jesus College, Oxford, (2007) Pigeons combine compass and landmark guidance in familiar route navigation. PNAS May 1 2007 104(18) 7471-7476
41. www.operationturtledove.org
42. Woodland Trust. www.woodlandtrust.org.uk
43. Tweetapedia. (No date) 21 Facts on Turtle Dove. Available at www.livingwithbirds.com/tweetapedia/21-facts-on-turtle-dove
44. North York Moors National Park. Available at www.northyorkmoors.org.uk
45. Bird identification guide: yellowhammer. Available at www.birdspot.co.uk/bird-identification/yellowhammer
46. Top 14 interesting facts about barn owls(No date) Available at www.gowiseowl.com
47. Barn Owl Trust. www.barnowltrust.org.uk
48. Lewis, D, (No date) Owls in mythology and culture, Available at www.owlpages.com
49. Swallow guide: migration, nesting and where to see. (No date) Available at www.discoverwildlife.com
50. Harrap, S, (2007) RSPB Pocket Guide to British Birds. Christopher Helm.
51. British Trust of Ornithology. (No date) BTO Birdfacts: House Martins. Available at app.bto.org
52. www.swift-conservation.org
53. Taylor, M and Warren-Chadd, R, (2016) Birds: myth, lore, and legend. Bloomsbury
54. www.springalive.com
55. Lederer, Dr R. (16.1.19) Swallows and Tattoos. Ornithology. Available at www.ornithology.com/swallows-and-tattoos
56. ID guide: Grasshoppers, crickets and groundhoppers. Available at www.bna-naturalists.org
57. Hadley,D. (27.8.20) The difference between grasshoppers and crickets. ThoughtCo, Available at www.thoughtco.com/difference-between-a-grasshopper-and-a-cricket-1968360
58. Grasshoppers and crickets (No date) Available at www.wildlifewatch.org.uk
59. Canal River Trust. (16.11.20) Waterway wildlife: Grasshopper. Available at www.canalrivertrust.org.uk/.../grasshoppers-waterway-wildlife
60. Fun grasshopper facts for kids. (No date) Available at https://sciencekids.co.nz/sciencefacts/animals/grasshoppers.html
61. Brigg, H, (24.10.19) Earthworms' place on earth mapped, Available at www.bbc.co.uk
62. Dann, L, Major survey finds worms are rare or absent from 40% of fields, Farmer's Weekly, (1.3.19) Available at www.fwi.co.uk
63. Agriculture Act 2020

64. Neil, P, (3.7.20) The crucial link between air pollution and biodiversity, Air Quality News Magazine, available at www.airqualitymews.com
65. www.coronationmeadows.org.uk
66. Plantlife, Hay festival? Action now for new species-rich grasslands (2018)
67. Barkham, P. Flower power: The movement to bring back Britain's beautiful meadows. The Guardian. (28.6.21) Available at www.theguardian.com
68. Ferguson, D. Pioneering rewilding project faces 'catastrophe' from plan for new houses. The Guardian. (21.3.21) Available at www.theguardian.com
69. Broughton, R, University of Oxford, (22.7.21) Monk's Wood Wilderness: Sixty years ago scientists let a farm field rewild- here's what happened. The Conversation. Available at www.theconversation.com
70. Barkham, P. UK Farmers to set aside 1% of the land for wildlife havens. The Guardian. (8.8.21) Available at www.theguardian.com
71. www.wildkenhill.co.uk
72. Richardson, R. (2017) Britain's Wild Flowers: A Treasury of Traditions, Superstitions, Remedies and Literature. Pavilion Books

UP HIGH AND HEATHS
1. WikiDiff (No date) Moorland v. Heathland- What's the difference? Available at https://wikidiff.com/moorland/heathland
2. Adams, Richard and Hooper, Max. (1978) Nature Day and Night. Kestrel Books.
3. North York Moors National Park. www.northyorkmoors.org
4. North York Moors National Park Authority Moorland Habitat Action Plan 2008-2012
5. Exmoor National Park. Available at www.exmoor-nationalpark.gov.uk
6. Slavik, Dr B. (1983) Wildflowers, Ferns and Grasses. Octopus
7. Plantlife. www.plantlife.org.uk
8. Podlech, D. (1996) Herbs and Healing Plants of Britain and Europe. Harper Collins.
9. Plants for a Future. Available at www.pfaf.org
10. Raven, S. (2011) Wild Flowers. The Observer
11. Parnassia palustris- Grass of Parnassus (No date) Available at https://first-nature-com/flowers/parnassia-palustris.php
12. Mabey, R. (2016) The Cabaret of Plants. Profile Books
13. Stafford, F. (2018) The Brief Life of Flowers. John Murray.
14. Wikipedia
15. Royal Horticultural Society. Available at www.rhs.org.uk
16. Whitehead, S. J. (1993). The morphology and physiology of moorland bracken and their implications for its control. Available at https://etheses.whiterose.ac.uk/2506
17. Bradley, D. (2012) Ptaquiloside – the poison in bracken. Bracken for breakfast? Available at www.chm.bris.ac.uk

18. Ravenscroft, J. (No date) Pace Egging: A Lancastrian Tradition. Available at www.timetravel-britain.com
19. Wildlife Trusts. www.widifetrusts.org
20. Juniper. Available at www.thejoyofplants.co.uk
21. www.ginfactory.com
22. Culpepper, N. (No date) Culpepper's Complete Herbal
23. Woodland Trust. www.woodlandtrust.org.uk
24. Stafford, F. (2016) The Long, long life of Trees. Yale
25. www.northpennines.org.uk
26. www.moorsforthefuture.org.uk
27. Irish Peatland Conservation Council. www.ipcc.ie
28. Peatlands and Soils. www.scottishwildlifetrust.org.uk
29. www.lexico.com
30. Ventress, G. (9.5.18) Lek it be: Black grouse mating season. Forestry and Land Scotland. Available at https://forestryandland.gov.scot/blog/black-grouse-mating.
31. RSPB www.rspb.org.uk
32. Moss, S. (2018) Mrs Moreau's Warbler. Faber
33. British Trust for Ornithology.(no date) BTO Birdfacts: Curlews. Available at app.bto.org
34. Barnes, S. The Meaning of Birds. (2016) Head of Zeus Ltd.
35. MacDonald Lockhart, J. (2016) Raptor: A journey through birds. 4th Estate.
36. Peregrine Falcons. www.discoveryofdesign2.com/peregrine-falcon-jet-engine
37. Bleckman, H; Brucker, C; Ponitz, B and Schmitz, A. Diving Flight: Aerodynamics of a Peregrine Falcon. PLOS ONE 9(2):e86506 doi. 10.137//journal.pone.0086506
38. BAE systems. (23.3.17) Research work on peregrine falcons inspires future aircraft technologies. Available at www.phys.org.
39. Hawk and Owl Trust. www.hawkandowltrust.org
40. Newsround (3.9.20) Rare hen harriers have a great year as 60 chicks are born. BBC. Available at www.bbc.co.uk
41. Top 10 fastest birds in the world (No date) Available at https://themysteriousworld.com/10-fas
42. Harding, A and Krestovnikoff, M. (2020) Birds. Bloomsbury
43. Interesting facts about the peregrine falcon. (No date) Available at www.justfunfacts.com
44. Carroll, S. E. (No date) Ancient and Medieval Falconry: Origins and functions in Medieval England. Available at www.r3.org
45. Ganninger, D. (23.4.22) Ten modern sayings that came from the world of falconry. Available at www.knowledgestew.com
46. Geddes, L. (29.11.11) Ravens use gestures to grab each other's attention. New Scientist. Available at www.newscientist.com
47. O'Hara, D, Ravens communicate better than most of the animal kingdom, Anchorage News 31.5.16 Available at www.adn.com

48. Nature. (2014) Ravens have social abilities previously only seen in humans. Nature Communications, 2014. DOI 10.1038/ncomms4679
49. 10 ravishing facts about ravens. Available at /www.mentalfloss.com/article/53295/10-fascinating-facts-about-ravens
50. Buehler, J. (19.2.20) Colour-changing animals, explained. National Geographic. Available at www.nationalgeographic.org
51. Earthrangers. (2.11.12) Polar bears have clear hair, so why do they look white? Available at www.earthrangers.com/risk
52. British Deer Society. Available at www.bds.org.uk
53. www.jncc.gov.uk
54. Grigson, G. (1962) The Shell Country Book. Dent.
55. 6 nightjar facts you need to know. (No date) Available at www.discoverwildlife.com
56. Tree, I. Wilding (2018) Picador.
57. Amphibian and Reptile Conservation Trust (2016) Slow Worm. Available at www.arc-trust.org/slow-worm
58. Springwatch BBC2. Broadcast 9.6.20
59. Change to smooth snake. Available at www.arc-trust.org/smooth-snake
60. Cumbrian Wildlife Trusts. (No date) Butterflies. Available at www.cumbrianwildlifetrust.org.uk
61. Cumbrian Biodiversity Centre Data 2016
62. www.ukmoths.org.uk
63. Devil's coach horse beetle. Available at www.wtf.nature.lifejournal.com/292302.html
64. www.ukbeetles.co.uk
65. Buglife www.buglife.org.uk
66. Case study: British dung beetles-here to help (15.4.19) Available at https://farmwildlife.info/2019/04/15/british-dung-beetles-here-to-help
67. Monibot, George, ITV 5.11.15
68. Climate Change Adoption Manual
69. Fix the Fells. (No dates) Available at www.fixthefells.co.uk
70. Wall, T. Burning Issue. The Guardian. (1.5.21) Available at www.theguardian.com
71. Exmoor Mires Partnership, Case study. Available at www.restorerivers.eu
72. Pennine PEATlife
73. Vinter, R. For peat's sake: the race is on to save Britain's disappearing moorland bogs. The Guardian (12.9.21) Available at www.theguardian.com
74. Hertfordshire Wildlife Trust www.hertswildlifetrust.org.uk
75. Great Heath Project. Dorset Wildlife Trust Available at www.dorsetwildlifetrust.org
76. www.eagleintroductionwales.com
77. Barkham, P. Return of the hen harrier. Unlikely protector helps revive endangered bird. The Guardian (4.12.21) Available at www.theguardian.com

78. Richardson, R. (2017) Britain's Wild Flowers: A Treasury of Traditions, Superstitions, Remedies and Literature. Pavilion Books

SIMPLY MESSING ABOUT IN BOATS
1. Freshwater Habitats Trust www.freshwaterhabitats.org.uk
2. Lewis-Stempel, J. (2019) Still Water: The deep life of the pond. Black Swan
3. RSPB www.rspb.org.uk
4. Moss, S. (2018) Mrs Moreau's Warbler. Faber
5. Colton, S. (8.4.17) Take on nature: Why the kingfisher is known as the 'halcyon bird.' The Irish News. Available at www.irishnews.com
6. Cocker, M. Country Diary: the much-misunderstood kingfisher. The Guardian (2.11.17) Available at www.theguardian.com
7. Wildlife Trusts www.wildlifetrusts.org
8. Eden, C. Things you might not know about the humble moorhen. Wildfowl and Wetland Trust (WWT). (16.6.20) Available at www.wwt.org.uk
9. Interesting facts about ducks. Available at www.justfunfacts.com
10. 10 Interesting facts about ducks. Available at www.thefactsite.com/duck-facts/
11. Countryside magazine. (14.10.20) Guide to Britain's geese species: how to identify, migration and where to see it. Available at www.countryfile.com
12. National Geographic (2018) 10 epic journeys of Britain's winter migrant birds. Available at www.nationalgeographic.co.uk/animals/2018/10/10-epic-journey-of-britains-wi...
13. Fowler, A. (2017) Hidden Nature: A voyage of discovery. Hodder and Stoughton.
14. Library of Congress. (No date) Why do geese fly in a V? Available at www.loc.gov/everyday-mysteries/browse-all-questions/item/why-do-geese-fly-in...
15. CBC Radio. Why do Canada geese honk while migrating? (24.8.19) Available at www.cbc.ca
16. Fun facts about geese. Available at https://opensanctuary.org/fun-facts-about-geese
17. Barnes, S. (2016) The Meaning of Birds. Head of Zeus Ltd.
18. www.swanlife.org.uk
19. www. wwt.org.uk
20. Wikipedia
21. www.merriam-webster.com
22. Scottish Wildlife Trust (No date) Osprey. Available at www.scottishwildlifetrust.org/species/osprey
23. Hawk and Owl Trust. www.hawkandowltrust.org
24. Davies, Professor N. (22.2.16) The reed warbler and the cuckoo: an escalating game of trickery and defence. University of Cambridge. Available at www.cam.ac.uk
25. Countryfile Magazine. (26.9.19) Britain's bittern population at a record high. Available at www.countryfile.com.

26. Pratchett, Dr M. (No date) Available at www.fashionfeathers.info
27. Harrabin, R. (22.4.20) Cranes make a comeback in Britain's wetlands. BBC. Available at www.bbc.co.uk
28. www.egretcraneproject.co.uk
29. www.froglife.org
30. Grigson, Geoffrey. (1962) The Shell Country Book. Dent.
31. Canal and River Trust. Available at www.canalandririvertrust.org.uk
32. Derwent, L. And Gillespie, T.H. (no date) Nature's Wonders. Collins.
33. Woodland Trust www.woodlandtrust.org.uk
34. Cohut, Dr M. (9/12/19) Researchers find over forty new species of fish in one lake. Medical News Today. Available at www.medicalnewstoday.com/article/327261
35. Essential fish facts you must know. (No date) Available at www.premier-fishing-tips.com/fish-facts.html
36. Carlson, S. M; Ceveria, I, Dias, E; Johnson, R. T; et al. (18.12.18) Scientists reveal secret migrations of fish. Available at www.frontiersin.org
37. How fish sense earth's magnetic field (2.5.17) Nature 545,9 (2017) doi://https://doi.org.10.1038/d41586-017-00633-7
38. Marine Scotland: Fishy Facts (25.11.19) www.gov.scot/publications/marine-scotland-fishy-facts
39. Trout Facts. Available at www.wildtrout.org/trout-facts
40. Mawle, G.W. and Milner, N.J. (2017) The return of salmon to cleaner rivers: England and Wales. Available at www.hwa.uk.com/site/wp-content/uploads/2017/12/Mawle-and-Milner-The-return-of-salmon-to cleaner-rivers.pdf.
41. Springwatch. BBC2. Broadcast 28.5.20
42. Fowler, A. (2017) Hidden Nature: A voyage of discovery. Hodder and Stoughton.
43. Ragged Robin- Not just a pretty flower: history and uses of the cuckoo flower. Herbs-Treat and Taste. (No date) Available at www.herbs-treatandtaste.blogspot.com
44. Lychnis flos-cuculi. Ragged Robin. Available at https://first-nature.com/flowers/lychnis-flos-cuculi.php
45. Plantlife. Available at www.plantlife.org.uk
46. Podlech, D. (1996) Herbs and Healing Plants of Britain and Europe. Harper Collins.
47. Raven, S, (2011) Wildflowers .The Observer.
48. History and meaning of the iris. (14.5.15) Available at www.proflowers.com
49. Wohlleben, P. (2018) The Secret Network of Nature. Vintage
50. Mabey, R. The Cabaret of Plants (2016) Profile Books
51. Stafford, F. (2016) The Long, long life of Trees. Yale
52. Hoffhines, A. and Miner, J. (No date) The discovery of aspirin's antithrombotic effects. Texas Heart Institute Journal 34(2):179-86. Available at www.researchgate.net/publication

53. Tree, I. Wilding (2018) Picador.
54. Scottish Invasive Species Initiative. American Mink. Available at www.invasivespecies.scot
55. Wild Otter Trust www.wildottertrust.org
56. www.scottishbeavers.org.uk
57. Freshwater Shrimp. Available at www.naturespot.org.uk/species/freshwater-shrimp
58. Freshwater Shrimp – Gammarus. Available at www.uksafari.com/freshwater_shrimps.htm
59. Greenstone, G(MD)(2010) The history of bloodletting. BC Medical Journal, Vol 52, No 1, Jan-Feb 2010 Available at www.bcmj.org
60. Martucci, J. (24.3.20) Medicinal leeches and where to find them: The rise and fall and resurrection of the humble leech. Available at www.sciencehistory.com
61. Isaac, S. (8.6.18) Why you should love a leech: bloodletting in microsurgery. Royal College of Surgeons England. Available at www.rcseng.ac.uk
62. Thomas, S. (July 2013.) Medicinal uses of terrestrial molluscs (slugs and snails) with particular reference to their role in the treatment of wounds and other skin lesions. World Wide Wounds. Available at www.worldwidewounds.com/2013/July/Thomas/slug-steve-thomas.html
63. Aitkin, R. (29.8.00) Snail slime 'could mend bones.' BBC. Available at www.bbc.co.uk
64. Gallagher, J. (27.7.17) Slimy snails inspire 'potentially lifesaving' medical glue. BBC. Available at www.bbc.co.uk /news/health-40730875.amp
65. Curtis, S. Snails are 'milked' for their slime use in cosmetics and beauty products. The Mirror. (22.7.19) Available at www.mirror.co.uk
66. Pitt, S. (14.11.19) Snail slime- the science behind molluscs as medicine. Available at www.theconversation.com
67. www.snail-world.com
68. Burrows, B. 9 Facts you never knew about snails: They have 14,000 teeth and they can kill you. The Mirror. (16.5.14) Available at www.mirror.co.uk
69. www.portugalfarmexperience.com
70. BBC. (23.8.13) Snails move 'faster than we thought', says study. Available at www.bbc.co.uk
71. www.snailracing.net
72. The Lakes with Simon Reeves. Broadcast on BBC2. (21.11.21)
73. National Trust. Riverlands
74. The River Trust https://therivertrust.org
75. Laville, S, McIntyre, N. (1.7.20) Water companies discharged raw sewage into England's rivers 200,000 times in 2019. The Guardian Available at www.theguardian.com
76. What is the difference between a signal crayfish and native crayfish? (10.11.17) https://waterways.org.uk/about-us/news/signal-crayfish
77. Williams, A. (4.6.19) Fear of killer shrimps could pose major threats to European rivers. University of Plymouth. Available at www.plymouth.ac.uk

78. Barkham, P. HS2 will destroy or damage hundreds of UK wildlife sites, says report. The Guardian. (15.1.20) Available at www.theguardian.com
79. North York Moors National Park.www.northyorkmoors.org.uk
80. Woodland on rivers boost in flood fight. Daily Mirror (25.9.21) Available at www.mirror.co.uk
81. www.wildennerdale.co.uk
82. Webb, J. A. (June 2020) Marsh lousewort as an ecosystem engineer in Oxfordshire fen restoration project. Available at www.freshwaterhabitats.org.uk
83. Anglesey and Llyn Fens LIFE Project. Available at https://naturalresources.wales/.../anglesey-and-llyn-fens-LIFE-project/?lang=en.
84. The Deep (Hull) (No date) Tansy Beetle. Available at www.thedeep.co.uk/tansy-beetle
85. Buglife www.buglife.org.uk
86. BBC. (7.5.21) Rare fish set for return to River Severn breeding ground. Available at www.bbc.co.uk
87. www.tweedforum.org/our-work/projects/river-till-restoration-strategy
88. www.nature.scot
89. UK Centre for Ecology and Hydrology www.ceh.ac.uk
90. Richardson, R. (2017) Britain's Wild Flowers: A Treasury of Traditions, Superstitions, Remedies and Literature. Pavilion Books
91. Amos, I. Call for volunteers to help birds devastated by avian flu outbreak in Southern Scotland. The Scotsman. (22.5.22) Available at www.scotsman.com
92. European Centre for Disease Prevention and Control. (No date) Questions and answers on avian influenza. Available at www.ecdc.europe.eu

THE DEEP BLUE SEA
1. Ordnance Survey
2. www.tides.today
3. www.ukcoastalguide.co.uk
4. The Guardian. Notes and Queries: Which British town is furthest from the sea? The Guardian (25.4.12) Available at www.theguardian.com
5. www.bbc.co.uk/weathercentre
6. National Tidal and Sea Level facility. (25.11.12) All about tides. Available at https://.ntslf.org/about-tides
7. Lye, H. (25.11.18) Five animals that can regenerate. The Guardian. Available at www.theguardian.com
8. Young Person's Trust for the Environment (No date) Life in a rock pool. Available at www.ypte.org.uk/factsheets/seashore/life-in-a-rock-pool
9. Countryside magazine. (31.5.18) Rockpool identification guide. Available at www.countryfile.com
10. Wikipedia

The Wonders of the Wild Places

11. University of Portsmouth. (18.2.15) Scientists find strongest natural material. Available at www.phys.org.dx.doi.org/10.1098/rsif.2014.1326
12. Wildlife Trusts www.wildlifetrusts.org
13. 11 supercool facts about crabs. (no date) Available at www.awesomeocean.com/news/super-cool-facts-crabs
14. Facts about crabs (No date)Available at www.factcity.com/facts-about-crabs
15. Barnacles (No date) Available at www.beachstuff.uk/barnacles.html
16. Field Studies Council. The Seashore (No date) Available at https://seashore.org.uk/theseashore/exposed rocky shore upper shore.html
17. National Ocean Service (NOAA) What are plankton? (No date) Available at www.oceanservice.noaa.gov / facts/plankton.html
18. Bittel, J. (17.11.21) See the microscopic world of plankton in stunning detail. National Geographic. Available at www.nationalgeographic.com
19. Plankton. Available at https://coast.noaa.gov/psc/sea/content/plankton.html
20. Sea kelp facts (31.7.19) Available at www.sciencing.com/facts-sea-kelp-4739866.html
21. Kelp facts for kids (No date) Available at www.kids.kiddle.co/kelp
22. Dybas, C. (8.6.18) Giant Kelp switches diet when key nutrient becomes scarce. Available at https://phys.org
23. Southsea Coastal Scheme
24. Buglife.www.buglife.org.uk
25. www.sussex.ac.uk/geography/researchprojects/BAC/biodiversity/shingleplants
26. Yellow horned poppy.www.norfolkwildlifetrust.org
27. Medicinal herbs: sea holly. Available at www.naturalmedicinalherbs.net
28. Plantlife. www.plantlife.org.uk
29. Raven, S, (2011) Wildflowers. The Observer
30. www.bumblebeeconservation.org
31. www.britishseafaring.co.uk
32. Clegg, B. (April 2021) Sixteen amazing facts about tardigrades, the world's toughest animal. Available at www.sciencefocus.com
33. Seashells. Available at https://coastalcare.org/educate/exploring-the-sand
34. Scales, H. (2015) Spirals in Time. Bloomsbury
35. De Dellker, Professor P; Prendergast, Dr A; Reeves. Dr J. (no date) Sea shells, sea shells. Available www.science.org.au
36. Wareham Oyster Festival. (2020) Oysters 101: Environmental benefits. Available at https://warehamoyster.com/2020/02/01/oysters-101-environmental-benefits
37. Wood, Dr J. (11.3.20) Are mussels and clams good for water quality like oysters? Chesapeake Bay Foundation. Available at www.cbf.org
38. Woolfe, S. (30.5.14) Why do spiral exist everywhere in nature? Available at www.samwoolfe.com

39. Palazzo, B. (27.6.16) The numbers in nature: the Fibonacci sequence. Available at www.eniscuola.net.
40. Gates, P. (10.7.19) UK sand dunes: best dunes to visit and wildlife to identify. Countryfile Magazine. Available at www.countryfile.com.
41. Podlech, D. (1996) Herbs and Healing Plants of Britain and Europe. Harper Collins.
42. Pavid, K. (14.7.16) Toxic talent of Britain's cyanide moths. Natural History Museum. Available at www.nhm.ac.uk
43. Tree, I. (2018) Wilding .Picador.
44. RSPB www.rspb.org.uk
45. Tegala, A. 7 facts you didn't know about terns. National Trust. Available at www.nationaltrust.org.uk
46. Moss, S. (2018) Mrs Moreau's Warbler. Faber
47. Sandwich tern. Available at https://jncc.co.uk/our.work/sandwich-tern...
48. Cleeves, T and Holden, P. (2006) The RSPB Handbook of British Birds. RSPB
49. World Atlas (27.4.17) The avian wonders: birds with the longest migration .Available at www.worldatlas.com/articles/the-avian-wonders-birds-with-the-longest-migration
50. Nicolson, A. (2017) A Seabird's Cry. William Collins
51. Collins Dictionary. www.collinsdictionary.com
52. Gill, V. (27.11.19) Great auk extinction: wiped out seabird. BBC. Available at www.bbc.co.uk
53. Galasso, S. (10.7.14) When the last great auk died it was by the crush of a fisherman's boot. Smithsonian Magazine. Available at www.smithsonianmagazine.com
54. Pavid, K. (No date) When worlds collide: the lessons of the great auk. Natural History Museum. Available at www.nhm.ac.uk
55. Schwartz, J. (11.9.18) Puffins are declining and climate change could become the largest cause. The Independent. Available at www.independent.co.uk
56. National Geographic Kids www.natgeokids.com
57. Hayward, I. Are seagulls endangered? RSPB. 20.8.10 Available at www.rspb.org.uk/blogs
58. Bawden, T. Numbers of urban seagulls in Britain has nearly quadrupled in the last 15 years, says research. The Independent. (31.7.15) Available at www.independent.co.uk
59. National Museum of American History, (No date) The Guano Trade, Available at www.americanhistory.si.edu
60. Kent, L. (6.8.20) Seabird poop is worth millions, says scientists trying to save birds. CNN. Available www.cnn.com
61. Super Interesting facts about seagulls you probably don't know (No date) Available at www.birdeden.com/interesting-facts-about-seagulls
62. History of featherbeds and duvets. (No date) available at www.oldandinteresting.com

63. Discover Wildlife. (No date) Available at www.discoverwildlife.com/animal-facts/birds/facts-about-oystercatchers
64. Probing with sandpipers (30.4.16) Available at www.birdnote.org
65. Smithsonian ocean. (No date) Just how big is the ocean? Available at https://ocean.si.edu
66. Marine Scotland: Fishy Facts (25.11.19) Available at www.gov.scot/publications
67. Abraham, J. Scientists study ocean absorption of human carbon pollution. The Guardian. 16.2.17 Available at www.theguardian.com.
68. Mashuda,S (17.12.21) In 2021 Scientists identified a new whale species in the Gulf of Mexico. Earth Justice. Available at www.earthjustice.org
69. Pavid, K. (No date) More whales and dolphins…in the UK. Natural History Museum. Available at www.nhm.ac.uk
70. www.uk.whales.org
71. www.us.whales.org
72. www.whalefacts.com
73. Pavid, K. (10.7.20) Echolocation gives whales lopsided heads. Natural History Museum. Available at www.nhm.ac.uk
74. Langley, L. (3.2.21) Echolocation is nature's built-in sonar. Here's how it works. National Geographic. Available at www.nationalgeographic.org
75. www.whale-and-dolphin-facts.com
76. Nidd, D. (17.2.21) The haunting music of whale song is an ocean of untapped seismic data, scientists say. Science. Available at www.sciencealert.com
77. Katz, B. (19.11.19) South Atlantic humpback whales have rebounded from the brink of extinction. Smithsonian Magazine. Available at www.smithsonianmag.com
78. Maxwell, D. (No date) Herman Melville. Britannica. Available at www.britannica.com
79. Cole, A. What is the difference between whales, dolphins, and porpoises? (25.11.20) Canadian Wildlife Federation. Available at www.blog.cwf-fcf.org
80. 19 killer whale facts about orcas. Available at www.factanimal.com/killer-whale
81. Grimm, D and Miller, G. (2020) Is a dolphin a person? Science magazine. Available at www.sciencemagazine.com,(21.2.20)
82. Thomas, T. Dolphins have similar personally traits to humans, study finds. The Guardian. (19.2.21) Available at www.theguardian.com
83. Striped dolphin factsheet. Cetacean Research and Rescue Unit. Available at www.crru.org.uk
84. Amazing facts about porpoises.www.onekindplanet.org
85. Autumnwatch BBC2. Broadcast 5.11.20
86. World Animal Protection. How plastic pollution is affecting seals and other marine life. (17.11.17) Available at www.worldanimalprotection.org
87. Pavid, K. (2.1.19) It's a record year for grey seal pups. Natural History Museum. Available at www.nhm.ac.uk

88. Pidd, H. Seal of approval: Farne Islands population boom gathers pace. The Guardian. (23.10.19) Available at www.theguardian.com
89. www. nature. scot
90. www.cornwallsealgroup.co.uk
90. Derwent, L. and Gillespie, T.H. (no date) Nature's Wonders. Collins.
91. www.mammals.org.uk
92. Amazing facts about grey seals. www.onekindplanet.org
93. Shaw, N and Abbott, C. (13.1.22) 6 types of jellyfish found off the UK-and what to do if you're stung. Available at www.cornwalllive.com
94. Marine Biological Association. (22.2.17) Fact sheet: Jellyfish. Available at www.mba.ac.uk
95. Jellyfish facts! www.nategeokids.com/sea-life
96. NHS. www.nhs.uk
97. Wolchover, N. (19.6.12) Does peeing on a jellyfish sting really work? Live Science. Available at www.livescience.com/34012-pee-jellyfish-sting.html
98. Smithsonian. (30.4.18) Bioluminescence. Available at https:// ocean.si.edu/ocean-life/fish
99. American Natural History Museum. (No date) How the jelly got its glow. Available at www.amnh.org
100. Pavid, K. (No date) Bioluminescence: Light in the dark. Natural History Museum. Available at www.nhm.ac.uk
101. Smith, Lauren. Glow in the dark sharks: new species discovered in Hawaii- and it glows. The Guardian. (19.12.17) Available at www.theguardian.com
102. Claes, Dr J. (21.2.13) Glowing sharks scares off predators with 'lightsabres.' Available at www.bbc.co.uk
103. Shark Trust
104. www.sharksider.com
105. Ten species you might not know live in our seas, World Wildlife Fund
106. Meeting the basking shark, RSPB. Available at www.rspb.org.uk/features
107. How sharks work. Available at https:// animals.howstuffworks.com
108. Collin, S.P. (2010) Electroreception in vertebrates and invertebrates. Encyclopaedia of Animal Behaviour. Available at www.sciencedirect.com
109. Conger, C. (No date) What is electroreception and how do sharks use it? Available at www.howstuffworks.com
110. Fish Science (No date) Fish colouration. Available at www.fishscience.co.uk/faq/fish-colouration
111. Rothschild, A. (2.4.13) Hagfish slime: the future of clothing. BBC. Available at www.bbc.co.uk
112. The history of herring fishing in the North Sea. (No date) Available at www.scarboroughmaritimeheritage.co.uk

113. Girl Museum (31.7.16) Herring girls Available at www.girlmuseum.org
114. Scottish Fisheries Museum (No date) The herring boom. Available at www.scotfishmuseum.org
115. Herring girls. www.mouthofthetweed.co.uk
116. Seahorse Trust. Available at www.theseahorsetrust.org
117. Hansford, D. (6.5.08) Cuttlefish change colour, shape shift to elude predators. National Geographic. Available at www.nationalgeographic.org
118. Reiter, S, Hulsdunk,P, Woo, T, Lauterbach, M.A. et al (2018) Elucidating the control and development of skin patterning in cuttlefish . Nature 562:361-366. http://dx.doi.org/10.1038/s41586-018-0591-3
119. University of Queensland. (21.9.16) Despite multi colour camouflage, cuttlefish and squid and octopus are colour-blind. Proceedings of the Royal Society B. Available at www.phys.org/pdf393657290.pdf
120. Monahan, P. (6.7.16) How 'colour-blind' cuttlefish may see living colour. Science. Available at science.org . DOI:10.1126/science.aaf5837
121. University of California Santa Barbara. (20.5.15) Study demonstrates that octopus skin possesses same cellular mechanism for detecting light as its eyes do. Journal of Experimental Biology. www.phys.org/news/2015-05-octopus-skin-cell...
122. Henry, L. (No date) Octopuses keep surprising us- here are eight examples how. Natural History Museum. Available at www.nhm.ac.uk
123. Starr, M. (3.3.21) A cephalopod has passed a cognitive test designed for human children. Available at www.sciencealert.com
124. Bradford, A. (8.6.17) Octopus Facts. Live Science. Available at www.livescience.com
125. Carnall, M. Why do cephalopods produce ink? And what's it made of anyway? The Guardian. (9.8.17) Available at www.theguardian.com
126. William, W.(No date) Kraken: The curious, exciting, and slightly disturbing science of squid. Harry N. Abrams
127. Planet Earth. BBC1. 24.1.21
128. Woody, T. (17.12.19) Huge amounts of greenhouse gases lurk in the oceans, and could make warming far worse. National Geographic. Available at www.nationalgeographic.com
129. www.oceanpreneur.com
130. Sustainable seafood. Available at www.worldwildlife.org/industries/sustainable seafood
131. Nicolas, A, (5.3.20) Meet the newly discovered ocean species: plastic. World Wildlife Fund. Available at www.wwf.org.uk
132. Amos, Jonathan. Ocean's extreme depths measured in precise detail. (11.5.21) The Guardian. Available at www.theguardian.com
133. Gill, V. (28.12.20) Atlantic discovery: 12 new species hiding in the deep. BBC. Available at www.bbc.co.uk

134. History of Plastics.www.plasticeurope.org
135. Woodford, C (16.5.22) Plastics. Available at www.explainthatstuff.com
136. Willis, L. (22.8.19)The most dangerous single source of ocean plastic that no-one wants to talk about. Sea Shepherd Global . Available at www.seashepherdglobal.org.
137. Greenpeace www.greenepeace.org.uk
138. National Trust available at www.naitonaltrust.co.uk
139. REV ocean. Available at www.weforum.org
140. Young Persons Trust for the Environment. (13.2.15) Oceans of plastic. Available at www.ypte.org.uk
141. UN Joint Group of Experts on the Scientific Aspects of Marine Pollution (GESAMP)
142. European Environmental Agency. Litter in our oceans. (2.6.14) Available at www.eea.europe.
143. Matei, A. Magnets, vacuums, tiny nets; the new fight against microplastics. The Guardian. (18.2.21) Available at www.theguardian.com
144. Denchak, M. (22.118) Ocean pollution: the dirty facts. Available at www.nrdc.org.
145. Seawatch Foundation
146. Countryfile. Broadcast on BBC1 29.8.21
147. Riopelle, J. (1.3.21) Increased noise pollution deafening marine life. Epigram. Available at www.epigram.org.uk
148. Ferrari, K. Understanding ocean noise pollution and its deadly impact on marine animals. IFAW. (20.5.20) Available at www.ifaw.org.
149. Carrington, D. Cacophony of human noise is hurting marine life, scientists warn. The Guardian. (4.2.21) Available at www.theguardian.com
150. How to stop overfishing. (No date) Available at www.theworldcounts.com.
151. Planet Earth. BBC1. 31.1.21
152. McRae, G. (16.4.20) Will climate change threaten the earth's other lung? The Revelator. Available at www.therevelator.com
153. Chu, J. (7.5.19) Phytoplankton decline coincides with warming temperatures over the last 150 years. Massachusetts Institute of Technology. Available at www.phys.org.
154. Palmer, B. (9.9.20) A massive surge in plankton has researchers pondering the future of the Arctic. Available at www.nrdc.org.
155. Virginia Institute of Marine Science (October 2008)
156. Nielsen, D and Petrou, Dr K. (26.8.19) Acid Oceans are shrinking plankton, fuelling faster climate change. University of Technology, Sydney. Available at www.theconversation.com
157. University of Leicester. (1.12.15) Falling phytoplankton failing oxygen: Global warming disaster could suffocate life on earth. Science Daily. Available at www.sciencedaily.com/releases/2015/12/151201094120.htm
158. University of California. (27.1.20) Oceanographers predict an increase in phytoplankton by 2100. Available at www.phys.org/pdf499329532

159. Carrere, M. (1.3.21) To fight climate change, save the whales, some scientists say. Available at www.mongabay.ca
160. Simms, D. (10.2.21) Professor of Marine Ecology at the University of Southampton. Ocean sharks and rays have decreased by 71% since the 1970s; here's what we must do. World Economic Forum. Available at www.weforum.org
161. Carrington, D. England plans ban on single-use plastic plates and cutlery. The Guardian (28.8.21) Available at www.theguardian.com
162. Arcanjo, M. (22.8.20) Plastic seas: navigating the waves of the ocean pollution. Climate Institute. Available at www.climate.org
163. Bawden, T. Carrier bag made out of egg shells to lead war on waste. i. (30.10.21) Available at www.inews.co.uk
164. Jenson, Nina. Chief Executive of REV ocean. www.weforum.org
165. Mavrokefalidis, D. (25.3.21) Teesside to become home to the world's first all plastic recycling plant. Available at www.futurenetzero.com
166. Grube, A. (2018) Daniel Burd: 17-year-old Eco-Expert. Available at www.oxsci.org.
167. Khanna, M. Scientists create an enzyme that can destroy plastic within days not years. India Times. (30.9.20) Available at www.indiatimes.com
168. Shukman, D. (7.9.18) Giant plastic catcher heads from Pacific Ocean clean-up. BBC. Available at www.bbc.com
169. www.theoceancleanup.com
170. Plymouth Chronicle. Innovative 'sea bin' will tackle plastic pollution. Plymouth Chronicle (26.4.19) Available at www.plymouthchronicle.co.uk
171. Patel, P. (26.3.20) The new kind of plastic is made to degrade in seawater. Anthropocene Magazine. Available at www.anthropocenemagazine.org
172. The plastic-free water bottle dissolves into nothing in less than three weeks (1.5.18) Available at https://bigthink.com
173. Kassman, A. (8.4.21) Ropeless fishing technology could help save rare whale, say scientists. The Guardian. Available at www.theguardian.com
174. McVeigh, K. Science is golden for whales as lockdown reduces ocean noise. The Guardian. (27.4.20) Available at www.theguardian.com
175. Polidoro, J. (23.2.21) A few fixes could cut noise pollution that hurts ocean animals. Scientific American. Available at www.scientificamerican.com
176. www.epigram.org.uk
177. Carrington, D. Cacophony of human noise is hurting marine life, scientists warn. The Guardian. (4.2.21) Available at www.theguardian.com
178. Ferrari, K (20.5.20) Understanding ocean noise and its deadly impact on marine animals. IFAW. Available at www.ifaw.org/uk/journal/interview-ocean-noise-pollution-impact
179. www.undp.org

180. Dunne, D. Fifty countries pledge to protect at least 30% of the world's land and oceans by 2030. The Independent (12.1.21) Available at www. independent.co.uk
181. McVeigh, K. Big days for British seas as bottom trawling ban in four protected areas proposed. The Guardian. (2.2.21) Available at www.theguardian.com
182. Barkham, P. Large expansion to 'blue belt' of UK's protected marine areas announced. The Guardian. (31.5.19)Available at www.theguardian.com
183. Hibbert, Tom. www.wildlifetrusts.org/blog/tom-hibbert
184. Springwatch. BBC. Broadcast 3.6.21
185. RSPB. (11.11.20) UK overseas territory becomes one of the biggest sanctuaries for wildlife. Available at www.rspb.org.uk
186. www.tristandacunha.org
187. 15 ways to prevent overfishing – solutions (22.8.17) Available at https://deepoceanfacts.com-ways-to-prevent-overfishing
188. How to stop overfishing. Available at www.theworldcounts.com.
189. Project Seagrass. Available at www.seagrass.org.uk
190. UN Environmental Programme. (1.11.19) Seagrass- secret weapon in the fight against global warming. Available at www.unep.org
191. Can seagrass help fight ocean acidification? (31.7.18) Available at www.phys.org.
192. The current state of seaweed (16.11.17) Available at www.aquaculturealliance.org.
193. Flannery, T. (31.1.17) How farming giant seaweed can feed fish and fix the climate. University of Melbourne. Available at www.theconversation.com
194. Golberg, A, Liberzon, A, Rubinsky B, Zollmann M. (2021) Multi-scale modelling of intensive macroalgae cultivation and marine nitrogen sequestration. Communications Biology (7.7.21) DOI 10.1038/s42003-021-02371 Available at www.nature.com/commsbio
195. www.seagrown.co.uk
196. Why Europe should be taking seaweed aquaculture seriously. (20.1.21) www.thefishsite.com
197. Renforth, P. (9.8.17) Preventing climate change by increasing ocean alkalinity. Available at www.eos.org
198. Plantlife- Dynamic Dunescapes.
199. Simons, P. Plantwatch: salt marshes are the unsung heroes saving our coastline. The Guardian (19.2.19) Available at www.theguardian.com
200. Milmo, Cahal. The great digital barrier reef. i. (30.10.21) Available at https://inews.co.uk
201. BBC. (25.8.19)Sharks and rays to be given new international protections. Available at www.bbc.co.uk
202. McCauley, D(19.2.21). A ten-step plan to save our seas. World Economic Forum. www.weforum.org /agenda/2021/02.a-10-s...

203. Richardson, R. (2017) Britain's Wild Flowers: A Treasury of Traditions, Superstitions, Remedies and Literature. Pavilion Books
204. Fava, M (9.5.22) how much of the sea has been explored? UNESCO. Available at www.oceanliteracy.unesco.org
205. Thompson, Dr D and Duck, C. Sea mammal research unit. Berwick and North Northumberland Coast European Marine site: grey seal populations status. www.dassh.ac.uk 103_4ea0d6d6e1a703.docx
206. Amazing Facts about lobsters . Available at https://onekindplanet.org
207. Lehnard, K (16.12.16) 63 Fun fish facts. Available at www.factretriever.com
208. 15 fun facts about octopus (12.12.20) https://factcity.com
209. National Geographic. (No date) Octopus Facts. Available at www.nationalgeographic.com
210. Bellis, M (31.7.21) A brief history of the invention of plastics. ThoughtCo. www.thoughtco/history-of-plastics-1992322
211. www.etymoline.com
212. National Geographic. Follow the Friendly Floatees. (No date) www.nationalgeographic.com
213. Pollution (no date) https://coastalcare.org
214. Plastic Pollution (no date) https://coastalcare.org
215. Shoemaker, S and Petre, A. (12.4.22) What is BPA? Should I be concerned about it? Healthline. Available at www.healthline.com
216. Clean up the world project. www.cleanuptheworld.org
217. Sala, E. (18.3.21) protecting the ocean is key to fighting climate change. Available at www.weforum.org/agenda/2021/03/protect-30-per-cent-ocean-and-biodiversity-climate-change

URBAN JUNGLE
1. www.statistica.com
2. Lambert, T. (14.3.21) A history of gardening. Available at https://localhistories.org/a-history-of-gardening
3. Britain Express. (No date) History of Gardens in the UK. Available at www.britainexpress.com
4. National Trust www.nationaltrust.org.uk
5. Wikipedia
6. www.gertrudejeykll.co.uk
7. Wildlife Trusts www.wildlifetrusts.org
8. Springwatch. Broadcast on BBC2 29.5.20
9. Moss, S. (2018) Mrs Moreau's Warbler. Faber
10. Robins. www.birdspot.co.uk/robins

11. 21 Facts about robins. Available at https://community.rspb.org.uk/wildlife/f/all-creatures/8957/21-facts-about-robins
12. Yong, E. (8.7.10) Robins can literally see magnetic fields, but only if their vision is sharp. Discover Magazine. Available at www.discovermagazine.com
13. Barnes, S. The Meaning of Birds. (2016) Head of Zeus Ltd.
14. Coal Tits. Happybeaks. (11.10.19) Available at www.blog.happybeaks.co.uk
15. Countryfile. BBC1. Broadcast 27.2.21
16. RSPB. (No date) The voices of Spring. Available at www.magpie.rspb.org.uk
17. Springwatch. BBC2. Broadcast 27.5.20
18. Woodland Trust www.woodlandtrust.org.uk
19. Hardman, S. (2017) The astonishing diversity of great tit song. Available at www.ogicablog.wordpress.com
20. Vaughan, A. Urban wildlife: when animals go wild in the city. The Guardian. (8.3.15) Available at www.theguardian.com
21. Smyth, R. (2017) A Sweet, wild note: What we hear when birds sing. Elliott and Thompson Ltd.
22. Country Living Magazine. (23.4.20) Good news! The dawn chorus is louder and clearer than it's been for a decade, thanks to lockdown. Available at www.countryliving.com
23. Porotsky, P. How do birds learn their songs? www.scienceline.org. (4.2.19)
24. Wade, H. (2010) The development of birdsong. Nature Education Knowledge 3(10):86
25. Countryfile Magazine (9.4.20) Guide to birds' nests: how to identify different species nests. Available at www.countryfile.com
26. Springwatch. BBC2. Broadcast 28.5.20
27. Walton, A.(2012) Home Sweet Home. RSPB Cymru Blog. Available at www.community.rspb.co.uk
28. Springwatch. BBC2. Broadcast 5.6.20
29. Stafford, F. (2018) The Brief Life of Flowers. John Murray.
30. Sasaki, J and Yamanouchi, K, Nagaki, M, Arima,H et al. (November 2015) Antibacterial effect of lavender flavour. Journal of Food Science and Engineering. DOI:10.17265/2159-5828/2015.02.006
31. Cavangh, H.M.A. and Moon, T. (2006) Antibacterial effect of essential oils, hydrosols, and plant extracts from Australian- grown lavender. International Journal of Aromatherapy. Vol 16. Issue 1. 2006. Pg 9-14
32. Do sunflowers follow the sun? (no date) Available at www.shesaidsunflower.com.
33. Pamela-Anne. 50 amazing facts about sunflowers. (no date) Available at www.shesaidsunflowers.com
34. Jones, L. (10.7.14) Undergraduate research studies sunflower power to clean up soil. University of Virginia. Available at www.phys.org.

35. Barnhart, M. (16.5.19) Cleaning up Chernobyl with sunflowers. Available at www.athenscienceobserver.com
36. Japan uses sunflower plants to decontaminate soil. www.scienceinfo.net
37. Sunflower does not reduce radioactivity in soil. www.scienceinfo.net
38. Interesting facts about daisies. (No date) Available at
39. Top ten facts about daisies(No date) Available at www.worldoffloweringplants.com
40. Ten facts about dandelions.(No date) Available at www.factfile.org
41. Facts about dandelions. (2015) Available at www.justfunfacts.com
42. www.dandeliondelights.com
43. Plantlife. Available at www.plantlife.co.uk
44. History of Roses (No date) Available at www.pickupflowers.com
45. Friends of the Earth (No date) 14 facts you need to know about bees. Available at https://friendsoftheearth.uk/nature
46. Friends of the Earth (25.7.17) Causes of bee decline. Available at https://friendsoftheearth.uk/nature
47. British Beekeepers Association. What is a bee? Available at www.bbka.org.uk/what-bee-is-this
48. Springwatch BBC2. Broadcast 11.6.20
49. Staughton, J. (22.1.22) How do bees find their way back to the hive? Available at www.scienceabc.com
50. Derwent, L. and Gillespie, T.H. (No date) Nature's Wonders. Collins.
51. Wohlleben, P. (2017) The Inner World of Animals. Vintage
52. Katy. (12.9.19) World's oldest honey. Save the Bees. Available at https://beemission.com/blogs/worlds-oldest-honey
53. Greenwood. D. (9.11.19) How is the queen bee chosen? Beehive Hero. Available at www.beehivehero.com
54. Hive hierarchy. (12.3.15) Available at www.beehivecollection.weebly.com
55. Hadley, D. (2020) "The Roles of Queens, Drones, and Worker Honey Bees." ThoughtCo, Aug. 27, 2020, Available at www.thoughtco.com/honey-bee-workers-drones-queens-1968099.
56. Greenwood. D. (30.5.19) Queen bees- a complete guide. Beehive Hero. Available at www.beehivehero.com
57. National Geographic for Kids (No date) 10 Facts about honey bees. Available at www.nategeokids.com /insects
58. How many eyes do bees have? (No date) Available www.busybeekeeping.com
59. Johnston, H. (27.2.17) Honey bees navigate using magnetic abdomens. Physics World. Available at www.physicsworld.com
60. www.mybeeline.co
61. World Wildlife Fund. Top ten bee facts. Available at www.wwf.org.uk
62. Howard, S, Dyer, A, Garcia, J. (6.2.17) Can bees do maths? Yes- new research shows they can add and subtract. Available at www.theconversation.com

63. Middleton, J. (17.2.21) Bees able to solve maths tests despite not understanding numbers, researchers find. The Independent. (17.2.21) Available at www. independent.co.uk
64. Westson, P. (5.6.19) Bees have human-like ability to link symbols to numbers, new study finds. The Independent. (17.2.21) Available at www. independent.co.uk
65. Hugo, K. (23.2.17) Intelligence test shows bees can learn to solve tasks from other bees. Science Magazine. Available at www.pbs.org.
66. Animal Einsteins. BBC2. Broadcast 21.2.21
67. Animal Einsteins. BBC2. Broadcast 28.2.21
68. Interesting facts about bees. (No date) www.justfunfacts.com
69. Aldiss, B. (24.8.18) Keep calm and wear white- how to avoid wasp stings. The Guardian. Available at www.theguardian.com
70. Hart, A. (7.9.15) 9 wasp facts that will get you buzzing about them. Available at www.sciencefocus.com.
71. RSPB www.rspb.org.uk
72. Perez, A. (2018) 7 mind boggling facts about parasitic wasps. Available at www.stillinfold.com
73. Pavid, K. (No date) Body snatchers eaten alive by parasitic wasps. Natural History Museum. Available at www.nhm.ac.uk
74. Jones, L. (16.3.15) 10 sinister parasites that control their hosts' movements. BBC. Available at www.bbc.co.uk
75. Le Page, M. (25.9.19) (Crypt-keeper wasps can control minds of seven species of wasps. www.newscientist.com Biology Letters DOI 1098/vsbl.2019.0428
76. Leslie, M. (30.4.18) How one parasitic wasp becomes the victim of another parasitic wasp. Science. Available at www.sciencemag.org
77. Hadley, D. (15.7.19) 10 Fascinating facts about ants. Available at www.thoughtco.com.
78. How ants communicate. (No date) Antkeepers. Available at www.antkeepers.com/facts/ants/communication
79. www.antkeepers.com
80. Caste System (No date) Antkeepers. Available at www.antkeepers.com/ facts/ants/caste-system-ant-societies/
81. Fun ant facts for kids (No date) www.sciencekids.co.nz
82. 11 amazing facts about ants (4.12.20) Available at www.somefactsabout.com
83. Wohlleben, P. (2018) The Secret Network of Nature. Vintage
84. Antkeepers (2021) Anatomy of Ants. Available at www.antkeepers.com
85. www.owlcation.com
86. Science Daily. (3.11.20) Ants swallow their own acid to protect themselves from germs. Available at www.sciencedaily.com.
87. Laffitte, M. (No date) Formic Acid. Available at chm.bris.ac.uk
88. Tragust, S, Herrman,C, Hafner, J, Braasch,R et al (2020) Formicine ants swallow their highly acidic poison for gut microbial selection and control. eLife, 2020, 9 DOI 10.7554/eLife.60287

89 Facts about caterpillars (March 2021) Available at www.justfunfacts.com
90 Bryant, M. Number of butterflies in the UK at a record low, survey finds. The Guardian. (7.10.21) Available at www.theguardian.com
91 Barkham, P. Britain's moths decline by about a third in 50 years, study finds. (3.3.21) The Guardian. Available at www.theguardian.com
92 www.wildlifeworld.com
93 Butterfly Conservation Trust. The State of UK's Butterflies (2015) Available at www.butterfly-conservation.org
94. Weston, P. (26.9.19) Huge moth that went extinct 50 years ago back in Britain. The Independent . Available at www.independent.co.uk
95. Johnson, S. (22.11.19) Facts about butterfly eggs. Science magazine Available at www.sciencemag.com
96. Hadley, D. (5.3.20) 10 fascinating facts about caterpillars. Available at www.thoughtco.com
97. Interesting facts about caterpillars (No date) www.justfunfacts.com
98. Interesting facts about caterpillars. (No date) Available at www.factsaboutbutterflies.net
99 Pavid, K. (14.7.16) Toxic talent of Britain's cyanide moths. Natural History Museum. Available at www.nhm.ac.uk
100. Fifty facts about caterpillars (No date) Available at www.everyfactever.com
101 Pavid, K. (No date) Butterflies: the science behind the colour. Natural History Museum. Available at www.nhm.ac.uk
102. Jomard, A. (26.3.21) "What Happens Inside the Chrysalis of a Butterfly?" sciencing.com, Available at https://sciencing.com/happens-inside-chrysalis-butterfly-8148799.html.
103. www.butterflyindentification.com
104. Jones, R. (No date) Can a butterfly defend itself in the chrysalis? Discover Wildlife Magazine. Available at www.doscoverwildlife.com
105. Vilazon, L. (No date) What's the powder on moth's wings? Available at www.sciencefocus.com
106. Hadley, D. (2020) "10 Fascinating Facts About moths" ThoughtCo, Aug. 27, 2020, Available at www.thoughtco.com/fascinating-facts-about-moths -1968179.
107. Newsround. (20.1.21) Butterflies: scientists explain how they fly. BBC. Available at www.bbc.co.uk/newsround/55729998
108. Nature. (2020) Physical and behavioural adaptions to prevent overheating in living wings of butterflies. Nat Comm. 11.551 (2020) htts://doi.org.10.1038/541467-020-14408-8
109. With butterfly wings there is more than meets the eye. (31.1.20) Available at www.sciencefriday.com
110. Hadley, D. (2020) "10 Fascinating Facts About butterflies." ThoughtCo, Aug. 28, 2020, Available at www. thoughtco.com/fascinating-facts-about-butterflies.

The Wonders of the Wild Places

111 Garcia de Jesus, E. (10.2.20) How thin, delicate butterfly wings keep from overheating. Science News. Available at www.sciencemag.org
112 Glick, M. (5.2.20) These infrared images show just how alive butterfly wings are. Available at www.popsci.com
113 Springwatch BBC2. Broadcast 9.6.20
114 Five fantastic facts about butterflies. St Nicks. (9.8.14) Available at www.stnicks.com
115 Henry, L and Teasedale, P. UK moths: 9 of the most colourful and distinctive. Natural History Museum. (No date) Available at www.nhm.ac.uk
116 Moths (no date) www.wildlifeworld.com
117 Arizona State University. Peppered moths: Natural selection- (No date) Available at www.asu.edu/peppered-moths-game/natural selection.html
118 Webb, J. (1.6.16)Famous peppered moth's dark secrets revealed. BBC. Available at www.bbc.co.uk
119 Hertzberg, R. (8.10.10) Why insects like moths are attracted to bright lights. National Geographic. Available at www.nationalgeographic.org
120 Lees, D and Zilli, A. (1.11.19) Why are moths attracted to light? Available at www.sciencefriday.com
121 Morell, V. (12.4.16) Your porchlight is causing moths to evolve. Science. Available at www.sciencemag.com
122 Tree, I. (2018) Wilding Picador.
123 Barkham, P. Experts solve the mystery of painted lady's winter disappearance. The Guardian. (19.10.12) Available at www.theguardian.com
124 Hadley, D. (2020) "10 Fascinating Facts About the Painted Lady Butterfly (Vanessa cardui)." ThoughtCo, Aug. 28, 2020, Available at www. thoughtco.com/facts-about-painted-lady-butterflies-1968172.
125 Eleven fascinating ladybird facts. (no date) Available at www.lovethegarden.com
126 Tregeur, P.(2018) The ladybird: Virgin Mary's beetle. Available at www.wordhistories.net
127 Our Lady's bug. (30.4.04) Available at www.catholicism.org.
128 Rolen, Father J.(No date) All about Mary. University of Dayton. Available at www.udayton.edu
129 www.english.stackexchange.com/questions/13352...
130 Amazing facts about ladybirds (no dates) www.onekindplanet.com
131 www.bbc.com/breathingspaces/ladybirds
132 The history of ladybug, ladybug. (23.8.17) Available at www.nurseryrhymesforbabies.com.
133 Smallest spider: identifying the 'itsy bitsy' spider in real life (18.11.21) Available at https://kidadl.com
134 Harris, S. (No date). 14 incredible spider facts. Discover Wildlife Magazine. Available at www.discoverwildlife.com.

135 Hadley, D. (2020) "Spider Silk Is Nature's Miracle Fiber." ThoughtCo, Oct. 29, 2020, Available at www. thoughtco.com/what-is-spider-silk-1968558.
136 Ashish. (20.1.17) What makes spider silk so special? Available at www.scienceabc.com
137 Human uses of the amazing spider silk: silk history and uses. (9.1.17) Available at www.pandasilk.com
138 Ferreira, Becky. Spiders can fly thousands of miles with electric power. Science News. (6.7.18) Available at www.resonancescience.org.
139 www.hedgehogstreet.org
140 Layton-Jones, Dr K. (No date) The history of public parks. Historic England. Available at www.katylaytonjones.com/publications
141 Ordnance Survey
142 Burford, Vanessa. (19.2.20) What is quintessentially a British park? Available at www.bbc.com
143 www.davidthorpe.info
144 Joseph Paxton www.parksandgardens.org / people/joseph.paxton
145 Mabey, R. The Cabaret of Plants (2016) Profile Books
146 Richard Spruce. Available at https://thedailygardener.org /otb20190910
147 York Philosophical Society www.ypsyork.org/resources/yorkshire-scientists-and-innovators/richard-spruce
148 Richard Spruce Collection-National History Museum. www.nhm.ac.uk
149 Ayto, J. (1990) Dictionary of word origins Bloomsbury
150 Newman, L. (8.9.19) Malaria could be eradicated by 2050, global health experts say. University of California, San Francisco. Available at www.ucsf.edu
151 Gordon, Lesley. (1983) Trees. Webb and Bower
152 www.churchofengland.org
153 Deak,B , Balazs, A, Loki, V, Lukacs, A et al (April 2019) Biodiversity potential of burial places-a review of flora and fauna of cemeteries and graveyards. Global Ecology and Conservation. Vol 18.e00614. Available at www.science-direct.com
154 New Forest National Park Authority. (No date)Bats. Available at www. newforestnpa.org.uk
155 www.bats.org.uk
156 The Wildlife Trust for Lancashire (No date) Bats. Available at www.lancswt.org.uk
157 Grigson, Geoffrey. (1962)The Shell Country Book Dent.
158 www.treesforlife.org
159 www.woodlands.co.uk
160 Barkham, P. (28.9.19) Britain's ancient yews: mystical, magnificent and unprotected. The Guardian. Available at www.theguardian.com
161 Stafford, F. (2016) The Long, long life of Trees. Yale
162 Animal Facts Encyclopaedia. Fox facts. (11.10.16) www.animalfactencylopaedia.com / fox-facts.html
163 DEFRA

164 Brinkhurst- Cuff, Charlie. Foxes surge into England's towns and cities. The Guardian. (16.4.17) Available at www.theguardian.com
165 Urban foxes, your questions answered. (no date) Available at www.bbcwildlife.org.uk/urban-fox
166 Osterloff, E. (No date) The secret life of urban foxes. Natural History Museum. Available at www.nhm.ac.uk
167 Red Fox-www.mammals.org.uk
168 Red fox- facts, diet and habitat information (No date) Available at https://animalcorner.org/animals/red-fox
169 www.foxproject.org.uk
170 Červený, Begall, Koubek, Novakova & Burda. 2010. Directional preference may enhance hunting accuracy in foraging foxes. Biology Letters http://dx.doi.org/10.1098/rsbl.2010.1145
171 Yong, E. (12.1.11) Foxes use the earth's magnetic field as a targeting system. Discover Magazine. Available at www.discovermagazine.com
172 Temming, M. (19.4.19) People may indeed have a sixth sense- for magnetism. Available at www.sciencenewsforstudents.org
173 Urban red foxes 'becoming more similar to domesticated dogs.' The Western Telegraph. (3.6.20) Available at www.thewesterntelegraph.co.uk
174 Vaughan, A. Urban wildlife: when animals go wild in the city. The Guardian. (8.3.15) Available at www.theguardian.com
175 Gagen, M. (2.2.21) Why keeping one mature tree is better for humans and nature than planting lots of new ones. The Conversation. Available at www.phys.org
176 Bubber, G. (31.5.18) The mysterious history of the London plane tree. Available at www.treesforcities.org
177 Vilazon, L (No date) . Do London planes actually absorb pollution into their bark? Available at www.sciencefocus.com
178 BBC. (10.5.04) Heatwaves can make trees pollute. Available at www.bbc.co.uk
179 Soriano, F. 22.12.17) How air pollution affects birds. Available at www.findingnature.co.uk
180 Field Studies Council (No date) Street and park trees. www.slnnr.org.uk
181 Woolf, J. Sycamore: colonist or custodian? The Hazel Tree (6.10.17) available at the www.thehazeltree.co.uk /2017/10/06/sycamore...
182 Urban wildlife. Natural History Museum. Available at www.nhm.ac.uk
183 Moss, S. Park life: the wildlife in Britain's cities. The Guardian (19.5.12) available at www.theguardian.com
184 King's College London. (No date)Study shows exposure to trees, the sky and birdsong in cities beneficial to mental wellbeing. Available at www.kcl.ac.uk
185 Acer Ecology. (No date)The benefits of urban wildlife. Available at www.acerecology.co.uk/urban-wildlife

186 British Ecological Society. (16.10.15) Making urban ecology count. Available at www.britishecologicalsociety.org
187 Coles, J. (31.7.15) How a city can save its wildlife. BBC Earth. Available at www.bbc.com
188 Johnson, K. (No date) Available at www.designingbuildings.co.uk
189 Stuart-Smith, S. (2020) The Well-Gardened Mind: Rediscovering Nature in the Modern World. William Collins
190 Folk ,E . How to stop light pollution from harming wildlife. (25.10.17) Available at www.wildlifearticles.co.uk
191 Friends of the Lake District. (No date) How light pollution impacts wildlife. Available at www.friendsofthelakedistrict.org.uk
192 Light pollution National Geographic . Available at www.nationalgeographic.org
193 Mandal, Dr A. (2019) Melatonin in mammals. News- Medical . Available at www.news-medical.net/health/Melatonin-in-Mammals.aspx
194 Rosane, O. (5.2.21) Traffic makes it harder for birds to think, scientists find. Eco Watch. Available at www.ecowatch.com
195 Neil, P, (3.7.20)The crucial link between air pollution and biodiversity, Air Quality News Magazine, available at www.airqualitymews.com
196 Laville, S. (6.8.20) Planning overhaul in England will damage nature, environmentalists warn. The Guardian. Available at www.theguardian.com
197 Ferguson, D. Pioneering rewilding project faces 'catastrophe' from plan for new homes. The Guardian. (21.3.12) Available at www.theguardian.com
198 National Trust (2020) New research shows the need for urban space. Available at www.nationaltrust.org.uk
199 Devon Wildlife Trust. Available at www.devonwildlifetrust.org
200 Weston, P. (4.12.20) Going wild? A radical green plan for Nottingham's unloved shopping centre. The Guardian. Available at www.theguardian.com
201 Avon Wildlife Trust. My Wild City. Available at www.avonwildlifetrust.org.uk
202 BBC. (2.2.21) Black poplar to feature alongside Bristol's new 'tiny forest.' Available at www.bbc.co.uk
203 Public Library of Science. (8.5.19) Urban trees 'live young, die fast' compared to rural forests. Available at www.phys.org
204 www.treesforcities.org
205 Briggs, H. (24.9.20) How lockdown birds sang to a different tune. Available at www.bbc.com
206 Newsround. (19.2.21) Climate change: How this hairy 'super plant' is helping fighting pollution. Available at www.bbc.com
207 Franklin-Cheung, A. (No date) Do trees reduce air pollution levels? Science Focus Magazine Available at www.sciencefocus.com
208 Traverso, V. (5.5.20) The best trees to reduce air pollution. BBC. Available at www.bbc.com

209 Myers, K. (29.10.18) Light pollution: The effects of artificial light on wildlife. Available at www.biofriendlyplanet.com
210 Plantlife's Every Flower Counts
211 www.naturehood.uk/wildlife.garden/build-a-pond
212 O'Boyle, C. Let's plant 1bn seeds. Daily Mirror (25.9.21) available at www.mirror.co.uk
213 Garner, D, (21.6.06) $13.6m research grant from Gates to develop plant to cure malaria. University of York. Available at www.york.ac.uk
214 Gunnerson, T. (7.5.22) Artemisin for cancer treatment. Available at www.webmd.com
215 National Biodiversity Network. (2.3.22) State of Britain's Hedgehog Report (2022). Available at https:// nbn.org.uk/news/state...
216 Interesting facts about wasps. Available at https://justfunfacts.com

HOPE FOR THE FUTURE?
1. Attenborough, Sir David. Extinction. Broadcast BBC1. 13.9.20
2. Countryfile. Broadcast BBC1. 24.10.20
3. Tree, I. (2018) Wilding Picador.
4. www.treesforlife.org

Bibliography

BOOKS

Hardman, I. (2020) The Natural Health Service. Atlantic

Hedgelink. (No date) Hedgerow Survey Handbook

Lewis-Stempel, J. (2019) The private life of the hare. Doubleday

Mabey, R. (2007) Beechcomings. Little Toller

Room, A. (2000) Brewster's Dictionary of Phrase and Fable . Cassell

Wulf, A. (2008) The Brother Gardeners: Botany, Empire and the birth of an obsession. Windmill Books

Articles

Albanese, J. How climate change skews phytoplankton blooms- and leads to fewer fish. Princeton University (24.7.20)

Barkham, P. 97% of Britain's wildflower meadows have gone: Here's why it matters. The Guardian (18.5.15)

BBC. (16.7.19) Smooth snakes: Bid to save UK's rarest reptile with £400,000 grant. Available at www.bbc.co.uk /news/uk-england-49000927

Calderwood, I. Nine brilliant ways the UK is cracking down on plastic pollution. www.globalcitizen.org (11.1.18)

Carrington, D. Future looks bleak for the soils than underpin life on Earth, warns UN. The Guardian. (5.12.20) Available at www.theguardian.com

Carrington, D and Wintow, P. UK will miss almost all its 2020 nature targets, says official report. The Guardian (22.3.19) Available at www.theguardian.com

Carrell, S. Scotland to ban mass culling of mountain hares. The Guardian. (18.6.20) Available at www.theguardian.com

Climate change shortens trees' lives. The Mirror. (9.9.20) Available at www.mirror.co.uk

Cobley, A. (5.4.21) Sir David Attenborough backs new technology that recycle all plastics. Available at www.goodnewsnetwork.org

Cockburn, H. Human birdsong boosts human well-being, study confirms. The Independent. (16.12.20) Available at www.independent.co.uk

Croxdale, J. P and Dias, M. P. (2019) Threats to seabird populations: A global assessment. Biological Conservation. Vol 237. Sept 19 525-537. Available at www.sciencedirect.com

Dewar, M. It's official – ivy's cool. Darlington and Stockton Times. (26.2.21). Available at www.darlingtonandstocktontimes.co.uk

Duarte, Carlos M, et al. (2020) Rebuilding marine life. Nature 580,39-51 (2020) Available at www.nature.com

Gabbatiss, J. Seabird populations have decreased by 70% as fishing competes for food. The Independent. (6.12.18) Available at www.independent.co.uk

Graham, I, (No date) Tackling malaria with fast-track plant breeding, University of York, Available at www.york.ac.uk

Greenfield, P and Weston, P. It's good for the soil: the mini rewilders restoring woodland. The Guardian. (23.5.20) Available at www.theguardian.com

Greenpeace.(No date) Plastic Debris in our Seas. Available at www.greenpeace.org.uk

The Wonders of the Wild Places

Hedge History. RSPB Available at www.rspb.org.uk

Hibbert, T. Wildlife Trust. (No date) Available at www.wildlifetrusts.org/blog/tom-hibbert

Higgitt, R. From cobwebs to silk: a world of human uses for spiders thread. The Guardian (3.10.16) Available at www.theguardian.com

James, V; Asmutis-Silvia, R; Ritter, F, Reyes, M.V. (January 2017) Whales- their future is our future. Available at www.researchgate.net/publications/ 327581...

Kleinman, Z. (20.5.10) Know your nettles: 10 bare facts. BBC. Available at www.bbc.com

Light pollution. www.nationalgeographic.org

Lindwall, C. (13.5.19) UN Report: A million extinctions and ecological collapse are on the way. Available at www.nrdc.org

Marshall, C. (3.10.19) More than a quarter of UK mammals face extinction. BBC. Available at www.bbc.co.uk

Marwood, C. Survey shows good signs for turtle doves. Gazette and Herald. (20.7.20) Available at www.gazetteherald.co.uk

Natural History Museum. (4.11.16) Dave the worm takes UK record for the largest size. Available at www.nhm.ac.uk

New Scientists. (8.4.15) Gimme the plane truth. New Scientists Issue 3016. Available at www.newscientists.com/lastword/mg2263016-700

Parkinson, J. How buzzards came to fly over the UK again. BBC News Magazine. (9.4.16) Available at www.bbc.co.uk

Rees-Warren, M. Why we should learn to love weeds. Darlington and Stockton Times (18.6.21) Available at www.darlingtonandstocktontimes.co.uk

Rewilding Britain. (No date) Rewilding and climate breakdown: How restoring nature can decarbonise the UK. Available at www.rewildingbritain.org

Ruggles, L. (2017) The minds of plants. Available at www.aeon.co

Science Daily, How birds can detect the earth's magnetic field, www.sciencedaily.com 6.4.18

Scott, D. (30.5.17) How the red fox adapted to life in our towns. The Conversation. Available at www.theconversation.com.

Simons, P. England's woodlands growing to 1000 per year. The Guardian. (22.11.01) Available at www.theguardian.com

The benefits of urban wildlife. www.acerecology.co.uk/urban-wildlife

Turretine, J. (12.10.18) Climate scientists to world: we only have 20 years before there's no turning back. NRDC. Available at www.nrdc.org.

University of Pennsylvania. (2.3.05) Why is the helix such a popular shape? Perhaps they are nature's space saves. Science Daily. Available at www.sciencedaily.com / releases/2005/02/050223135535.htm

Waddington, M. The important of seagrass: why we should conserve the marine habitat. www.gui.co.uk (2020)

Walker, P. Boris Johnson unveils £1.2bn for climate and endangered species. The Guardian. (22.9.19) Available at www.theguardian.com

Weston, P. Bees have human-like ability to link symbols to numbers, new study finds. The Independent. (5.6.19) Available at www.independent.co.uk

Why Europe should be taking seaweed aquaculture seriously. The Fish Site. (20.1.21) Available at www.thefishsite.com

Winterman, D. (8.3.15) The surprising uses of birdsong. BBC. Available at www.bbc.com

REPORTS

Biodiversity 2020: A Strategy for England's wildlife and ecosystems. (19.8.11) Available at www.gov.uk

British Ecological Society. (2.6.11) UK National Ecosystem Assessment (2011). Available at www.britishecologicalsociety.org/the-uk-n...

Coalition Clean Baltic. (January 2020) The Blue Manifesto: A road map to a healthy ocean in 2030. Available at www.ccb.se/blue-manifesto

Future of the sea: the future of marine biodiversity. (2017) UK government. Available at www.gov.uk

Jncc. Seabird populations: threats and causes of change 1986-2015. Available at www.jncc.gov.uk

Upland Management Manual. Natural England. (1.2.01) Available at www.publications.naturalengland.org.uk/publication/82...

Woodland Trust. State of the UK's woods and trees. (2021) Available at

www.woodlandtrust.go.uk/state-of-uk-woods-and-trees

Websites
www.2.gov.scot
www.bbsrc.ukri.org
www.blackthornandbone.co.uk
www.bluecross.org.uk
www.dkfindout.com
www.growwilduk.com
www.kidskonnect.com
www.kingsfund.org.uk
www.moorlandassociation.org
www.nfu.online.com
www.onegreenplanet.org
www.phrases.org.uk
www.riverfosssociety.co.uk
www.scottishraptorsstudygroup.org
www.seethewild.org
www.soilassociation.org
www.thebritishbirds.com
www.thecirculateiniative.org
www.theecologist.com
www.theenglishgarden.co.uk
www.thefield.co.uk
www.thespruce.com
www.transceltic.com

ABOUT THE AUTHOR

Rachel Lister-Jones lives in her home county of North Yorkshire, just south of the North York Moors National Park, with her whippet, Lady Millicent. She has been fascinated with nature since a very young age – her first word was 'bird' – and she is particularly interested in botany and ornithology. She loves walking in North Yorkshire and the Lake District, and runs a mindful nature walks company (@wildwalksyorks) in North Yorkshire, as well as volunteering for the Woodland Trust and her local environmental group. She is also a poet and writer and is very inspired by the natural world.